The Selborne Pioneer

Gilbert White as Naturalist and Scientist, a Re-examination

Ted Dadswell

CENTAUR PRESS
an imprint of OPEN GATE PRESS
LONDON

This revised edition first published in 2006 by Centaur Press,
an imprint of Open Gate Press,
51 Achilles Road, London NW6 1DZ.
Reprinted in 2009.

Original edition published in 2002 by Ashgate Publishing.

British Library Cataloguing-in-Publication Programme
A catalogue reference for this book is available from the
British Library.

ISBN: 0 900001 56 9
EAN: 978 0 900001 56 7

Printed and bound in Great Britain
by ImprintDigital.net.

for

Anne Hamilton

Ted Dadswell grew up on a Sussex farm and was educated at Collyer's School, Horsham, and Durham University. In 1978, after a period of teaching in the North East, he moved to London, since when he has worked as a freelance historian and editor. He now lives in East Oxford.

Reviews of the hardback edition:

Digests with a multiplicity of reference sources easily become a word jungle with little lucid flow. Ted Dadswell expertly avoids this problem, and although his book reflects in part a personal journey, his own views embelish an intriguing analysis, assisting in producing a fascinating and readable text. Recommended equally to Selborne devotees and to newcomers.

Jim Flegg, *Ibis*.

From earthworms to earth tremors, The Selborne Pioneer frames the practice of natural history within the rural and social structures that guided White's observations when he exchanged letters during the winter and conversed with neighbors on local lanes and paths during the summer.

M.D. Eddy, *Isis* (USA).

This is a valuable addition to the bookshelf of anyone with an interest in the remarkable curate of Selborne.

Selborne Society, *Newsletter.*

CONTENTS

LIST OF ILLUSTRATIONS

ACKNOWLEDGEMENTS

The author is grateful to the publishers of the following works for permission to use extracts. Paul G. M. Foster, 'The Gibraltar Correspondence of Gilbert White', *Notes and Queries* (Oxford University Press, 1985), and Humphry Primatt, *The Duty of Mercy and Sin of Cruelty to Brute Animals,* ed. Richard D. Ryder (Centaur Press, 1992). Also to the British Library for permission to reproduce a letter from Gilbert White to Thomas Pennant dated November 6th 1767, and two pages from White's *Naturalist's Journal* for September 1789.

PREFACE

Gilbert White (1720–93) is extraordinary in a number of respects. For one thing, and although *The Natural History of Selborne* (1789) has been regularly re-issued during the past two hundred years, we can still encounter him first by way of what might be called the Gilbert White mythology. This 'mythology', a combination of stories and assumptions, is often quite at variance with what White did and wrote. It represents him as naïve if also well meaning, childlike and sometimes childish.

The continuing effect of the mythology can easily be illustrated. In February 1984 the broadcaster Stefan Buczacki remarked on the air that 'The old naturalist Gilbert White hated earthworms and wanted to get rid of them, unlike Darwin who wrote a book in their defence'.[1] This is untrue and, more than this, is the opposite of the truth; in a short but concentrated paper – published ninety-three years before Darwin's work appeared – White insisted on the ecological value of earthworms.[2] But Buczacki's formulation sounded right, to the speaker himself and presumably to the other members of the Gardener's Question-Time panel. He was not claiming to have studied White (sadly, for White, too, was an enthusiastic gardener), but if the 'old naturalist' had said something about earthworms, it was likely to have been quaint or at any rate unsophisticated.

White was a contented man, unambitious and more than this 'a bit lazy', we are sometimes told. In old age, when he had left a letter unanswered for some months, he described himself as a 'procrastinator', and procrastination is often assumed to have typified him. In 1979, *Selborne* itself was described by a well received author as 'the testament of static man'.[3] 'Static', but also idiosyncratic to the last. On several occasions, White tried to get responses from seemingly phlegmatic creatures by shouting at them through a megaphone, a 'speaking trumpet'. Or a trumpet of some sort. That he *played* the trumpet to various animals to test their hearing, has been confidently passed on by several writers.[4] The idea of a trumpet-playing White is amusing – and is perhaps not entirely foreign to aspects of White's character – but it is, virtually, without historical basis.

The mythology can distort facts, and just as importantly can obscure facts. It has sometimes been implied in the course of the last two or three decades that if Gilbert White still stands out from his contemporaries, it is as a literary figure, rather than as a 'naturalist and scientist'. In the words of Keith Thomas, were it not for White's 'special sensibilities and literary skills' we should have little reason to distinguish him today from numerous of the other amateur naturalists of his time.[5] We can be grateful to Professor Thomas for reminding us of how much recreational nature study went on in eighteenth century Britain – and for insisting that Alexander Pope was not alone in protesting against 'barbarity to animals'. But if the case made out in this book is justified, to describe White as less than a rigorous and highly original naturalist is, again, to quite misrepresent both the man and his work.[6]

White was a gifted and an unspoilt stylist; here there is no need for disagreement (almost any paragraph of *Selborne* bears out the point). But he was also an amateur natural historian, and one – to begin to correct some of the misapprehensions – with a shrewd understanding of much that the specialist and professional naturalists of his time were doing. A valuable aid to the assessment of White in his scientific role was the editing and transcription in full of his 'Gibraltar correspondence', in 1985. These are letters written by White to his younger brother John from 1769 to 1772 inclusive, while the latter was chaplain to the garrison at Gibraltar. Encouraged by Gilbert, John White was trying to produce a natural history of Gibraltar, and was sending miscellaneous consignments of specimens to England. In the letters, Gilbert is giving his brother advice and instruction, and is reporting on the identification and classifying of the specimens which he, Gilbert, was organising and himself contributing to. The brisk tone of the letters, and White's well informed approach to this supervisory work, tell their own story. The extract provided below is from a letter to John White dated January 25, 1771. The detail would need interpretation but it is enough to mention here that Geoffroy was the author of a new French entomology, a work replete with clear and accurate engravings, and Brisson was a French ornithologist. Scopoli – as well as Linnaeus – was an established naturalist and taxonomist, and John Ray was an earlier, ground-breaking authority in both these roles.

> I received your kind letter of Oct: 19: & wrote you an answer on Nov: 6. . . . The merula passer inditarius of Ray is said to be a fine songster . . . Y^{r} winter swallow is undoubtedly the hirundo rupestris of Scopoli: he has found out their summer residence, but you will have the merit of discovering their winter quarters. Brisson mentions a tridactyl quail from Madagascar . . . His ornithology is extravagantly dear: 7, or 8 guineas. Geoffroy will set you right, by means of his cuts, in many genera of insects.
>
> Now Mr Twisse is returned be sure to get his conjectures on ye currents of the streights . . . Have you not in Spain some crown-flocks of sheep which migrate with the seasons from N: to S: get some anecdotes about them. Mr Pennant makes his artist take all y^{r} most curious birds; & promises the drawings shall be forth-coming if wanted to engrave from. Describe the vulture percn: most minutely; & learn, if you have an opportunity, how the sexes differ. Get the skin of the Lupus aureus from Barbary, & describe it well. Scopoli's new hirundo alpina is nothing, I think, but the hirundo melba; which is indeed a noble swift . . . Write to Scopoli . . . ask him as gravely as you can how he is sure that the woodcock, when pursued, carries off her young in her bill. . . . Scopoli's Icones will prob: disappoint you . . . Geoffroy's are the best I have seen. The bird you call a parus (if it be not a blackcap; I declare I am not sure) is a nondescript: if new, call it motacilla atricapilloides: Mr Pennant thinks it new. Y^{r} purple-winged vespa is no doubt the crabroni congener &c of Ray; & if you can find that it has "thorax ad latera postice utrinque dente-notatus," I shall acknowledge it to be ye sphex bidens Linn: I am sending all y^{r} insects to Nep: Ben Wh: in town, & shall get Mr Lee of Hammersmith . . . to inspect & ascertain them. . . . You will now be able to measure the rain of y^{r} climate: the mean quantity of rain p^{r} ann: in Rutland is 20 3/4 in. . . . The best foreign faunists are very barren in the manners of y^{ir} animals. What do the panorpae coae do with those enormous remiform wings? send a better specimen of the plant they haunt. But do they not still belong to the water? Get some account of the prickly heat or fever: & the exact height of y^{r} mountain.[7]

And so on. White is moving from birds to mammals, and then from birds to insects, here. (Regarding the *Panorpa coa*, a kind of lacewing, he suggests in another letter that its exaggerated wings 'relate somehow to mating'.) Insects, he willingly accepts, are not his strong subject; he was always principally a watcher and investigator of birds. But the confusion which could still accompany the discovery of a 'new' bird is clear from the extract, and because there might be several unclassified species in any of John's boxes and packages, White passes on all the specimens, including the birds, to recognised experts. The correspondence alerts us, though, to White's own intellectual range and keenness. He is not over-confident, but he is clearly quite familiar with the requirements of the experts, and his brother, the trainee, is to be allowed no slackness or short cuts.[8]

The Gibraltar study – John White's study – reached an advanced stage, we know, but was never published and the manuscript has been lost. As we can tentatively re-construct it from the correspondence and from various comments in *Selborne*, it would have been a considerable advance on the local 'natural histories' of the day. These were in some cases 'barren' not only in 'manners' but in species. As I have put it in Chapter Five below, the British examples were often more concerned with human entrepreneurs, and how they could fit into and exploit a locality, than with the challenges the same locality offered to non-humans.

The Gibraltar study would have been animal- and plant-centred, and would have been interspersed with remarks on manners. But the Gibraltar correspondence, while it illustrates the application White was capable of, introduces us only incidentally to his own characteristic interests. Gibraltar was a 'nondescript' locality. In White's efforts to help his brother, he buckled down to identification and even taxonomy. But he, Gilbert, was not a natural taxonomist. In taking the Selborne parish as his own scene of operations – and in concentrating even here on common creatures – he was effectively turning away from concerns of this kind. What then was central to the 'natural knowledge' we can properly associate with him? What would be the main subject matter of *The Natural History of Selborne*? We are provided with pointers as his personal prejudices occasionally appear in the Gibraltar letters: 'True naturalists', he insists at one point, 'will thank you more for the life & conversation of a few animals well studyed & investigated; than for a long barren list of half the Fauna of the globe.'[9]

'Life and conversation', or behaviour in a social and physical habitat, was White's special forte; in his own work, he studies *non-human behaviour* scientifically. White was not the innocent, or even the entirely ingenuous, naturalist he sometimes seems – and in even the *Selborne* papers, he from time to time takes part in 'systematics'; but he was drawn to animal and plant behaviour. With his talents, and in particular I have suggested his methodological aptitude, he could embark on behavioural studies very conveniently in a Hampshire parish. Even in a Hampshire parish, he thought, he could be one of a widely scattered team of researchers contributing jointly to the investigation of various behavioural 'puzzles'.

The behavioural emphasis is what gives *Selborne* its peculiar place in the history of natural history studies. (White's claim concerning what 'true naturalists' value is surprising, as an eighteenth century claim.) He hoped for the emergence of a community of behavioural workers, but in his time this did not happen. Although the idea of such a community is familiar enough today, organised, widespread 'habits and habitats' research was not developed by White's immediate successors; in Britain at any rate, it came about only in the twentieth century. At the time of the First World War, when the young Julian Huxley began to take outdoor behaviour seriously, what he calls 'live ecology' was still 'not fully approved of by "classical" zoologists'.[10] Even in 1936, the naturalist and writer Fraser Darling could complain that 'The great bulk of papers on animal behaviour lift the organism from its normal environment and place it in a set of artificial conditions, and this often results in findings which are not valid for the interpretation of representative behaviour.'[11]

If this is true, White was an innovator, and one comparable to Gregor Mendel, the geneticist, say, or Charles Babbage the nineteenth century designer of ambitious computers. Several of the problems he was involved with concerned 'birds of passage', and even when not dealing with these travelling species, he, too, was opposed to lifting animals from their normal environments. But he was observational and, in a sense, also experimental. We have sometimes been told that he was a fugitive if not from science, then from rigorous science, but this notion too, it can be shown, belongs with the Selborne mythology.

According to the philosopher George Berkely, 'a man must not only demonstrate the truth, he must also vindicate it against scruples & establish'd opinions which contradict it.'[12] I have tried to act on Berkeley's advice, or warning, in this book. In addition to *The Natural History of Selborne* and the extant private letters, I have made extensive use of White's manuscript journals and notebooks. He begins to take on his proper stature 'as naturalist and scientist' only when *Selborne* and the lengthy *Naturalist's Journal* are considered as distinct, complementary works. (Despite a view often accepted by commentators, the *Naturalist's Journal* is not merely the *Natural History of Selborne* in unedited form.)[13] White seldom intruded physically on the creatures he studied (his garden work excluded), and for him any animal or plant had an ecological dimension, as we would now term it, but he saw that if his records and readings were accurate, the handling of *these records and readings* could give him a means of investigating his behaving subjects rigorously. It was not experiment but inappropriate experiment that he avoided and challenged by implication; and the procedure he approved and tried to use most often, a comparative testing and eliminating by appeal to behavioural records, was especially well suited to the study of animals and plants, or the study, rather, of their 'life and conversation'.

White's 'comparative' aspect has generally been neglected, and present-day historians of science are uncomfortable with the description of a scientist as 'ahead of his time'. He has sometimes been, so to say, squeezed into what was happening around him. However, to arrive at a worthwhile assessment of the Selborne naturalist we have to ask historical as well as scientific questions; we must know what he himself was part of, culturally and psychologically, as well as whether his findings have subsequently been tested and verified. Comparative analysis using recorded data has often been crucial to the activities of twentieth century naturalists and biologists, but White was the contemporary of Thomas Gainsborough and the near contemporary of Samuel Johnson. As he can be found tutoring his brother John, he is not an altogether surprising figure; his intellectual grasp may be unexpected, but the brothers were setting up what is immediately recognisable as an eighteenth century project. But where did White the 'behavioural scientist' come from? How historically speaking is this Gilbert White to be accounted for?

A scientific assessment does not rule out an historical examination, or vice versa, but little serious attention has been given over the years to White's cultural background. (His life and work, indeed, have sometimes been regarded as an almost magical occurrence, requiring no explanation.)[14] In fact, and while his observing and analysing were exceptional, it is not difficult to find at least some of the influences concerned in them. We should avoid 'bland, confident general statements about whole groups of men, or classes or nations'.[15] In White's case, the relevant influences are not to be found in the twentieth century, fairly obviously, but neither are they all to be found in the second half of his own century. He was considerably affected by an Augustan 'science of man', I have suggested. He worked within a Baconian, or more satisfactorily, a 'Newtonian and Lockean' tradition, a tradition encouraging observation, a 'spirit of accuracy' and what were called 'comparisons of ideas'. As it turns out, White foreshadowed some much later work. The history of culture can sometimes take a spiralling or helical course; within the science of the past three and a half centuries, anyway, different sources have often produced compatible results. If I am right, White and the more recent 'outdoor' naturalists speak the same language – whether or not they use the same words.

I have tried to define my own terms fairly carefully in what follows. As I understand it, 'science' is a project as well as a body of results, and scientific questions are about events, or what happens, in the phenomenal or natural world. More precisely, they are about reliable time and place relations in or between phenomenal events, and thus about the making of provisional predictions. The scientist tests provisional predictions, and does so with a use of controls. Because his materials are quantifiable, he can – and in a sense inevitably does – share his work with other researchers. But this is by no means an elementary sort of science, then, calculated to accommodate all comers. It does not encourage the timid or unenterprising researcher, even if, as is certainly the case, it does not restrict the scientist merely to work within a laboratory.

There will be those who object on principle to my finding similarities between White and the present-day behavioural investigator. This book was not begun, or written, as a polemical exercise, but a degree of polemicism has been unavoidable. The three preoccupations which brought me to White and his journals were eighteenth century British thought and culture, animal psychology and bird study. When I began planning the book, some fifteen years ago, I was more aware of White's twentieth century supporters than of those who have thrown doubt on his achievement. (The commentators referred to at the beginning of this preface were by no means the only twentieth century writers on White.) What I wanted to say about the Selborne naturalist was new in part, but I thought of it as augmenting what had already been done.

Various of these 'supporters' were themselves experienced naturalists. They included James Fisher, Ronald Lockley, Max Nicholson and Walter Johnson. In general, their contributions appeared earlier than, and differed widely from, the views touched on above. In his introduction to an edition of *Selborne*, Max Nicholson described White as 'the first herald of the dynamic view of Nature'.[16] 'First' may or may not be correct, but as I am presenting Gilbert White, he described an organic living world and could find adaptability in a great range of living things. In his book on White, Walter Johnson refers to his subject's 'strange, perhaps unique, position as a forerunner'.[17] Johnson was writing in the late 1920s, and much has happened in behavioural natural history since that time; but he was, if anything, more prescient here than he realised.

James Fisher thought that quite enough had been heard over the years of the 'charm' of Gilbert White, and regarding *Selborne,* he insists that 'its scientific value is as important as its general atmosphere'.[18] Walter Johnson acknowledges the charm, but he points also to White's 'simple fearlessness and impartiality'.[19] White cannot be summed up in one bold statement; the Gilbert White supporters have sometimes subscribed to their own mythology; but his approach was simultaneously restricting and liberating. With behaviour as his subject matter, he avoided the 'mechanical' biology of many of his contemporaries, and examined responses which were quite unrecognised by the mechanists.[20] He thought about migration, but also 'territory' and 'dispersion' in birds; he drew attention to the puzzle, as he calls it, of closely related species adapted to widely differing habitats.[21] In addition, he saw that behavioural evidence could be precise and was communicable. He was, if not an unwilling innovator, then an unwilling individualist; in the process of demonstrating that several observers were better than one, he from time to time almost invented co-workers. He offered his services as a team member although the teams had still to appear; within a few years, he thought, behavioural nature study would gain the necessary recruits. He was mistaken in this. But more than two hundred years later, we can co-operate with him. Crucially, if I am right, we can re-run and evaluate the enquiries he began.

But *The Natural History of Selborne* has appeared to date in more than two hundred editions, in English alone. Is a further book on White as naturalist really required? Yes, because whole aspects of White's 'natural knowledge' were obscured during the nineteenth and early twentieth centuries and are still being re-discovered – and because these aspects enhance the seeming anomaly of Gilbert White, or render it more challenging. Whether or not we can wholly account for him, White was both a pioneer and what I have called a 'perennial' naturalist. He not only 'started us all bird-watching', in the words of James Fisher, but is still a source of factual information and a methodological paradigm. His influence is difficult to ascertain – and in the comfort of a jet airliner we can easily underestimate the achievements of the first fliers; but on the evidence summarised in these chapters, he was an early and quite extraordinary exponent of modern behavioural biology.

I owe Walter Johnson a special debt. Johnson was unfortunate in his timing; the twentieth century work with which White has a particular affinity was in only its early, tentative stages when Johnson was writing; but his study, *Gilbert White, Pioneer, Poet and Stylist* (1928), was the best book-length twentieth century guide to White's various activities. I have made regular use of it.[22] While writing my own book, I have also turned repeatedly to Paul Foster's, *Gilbert White and his Records* (1988).

Books and papers in the Selborne Collection were kindly made available to me by the Committee of the Selborne Society, and subsequently the Ealing Central Library, when I began work. I have been grateful for the enormous resources of the Bodleian and Radcliffe Science libraries at Oxford, and was glad to have access to the manuscript of the *Naturalist's Journal* in the British Library.

I want to express my thanks to various good-natured readers and advisers. The late Colin Tubbs, ecologist and author, read the book at an early stage and corrected various mistakes. Professor Aubrey Manning of Edinburgh, Dr Marian Stamp Dawkins of Oxford, and Professors David Knight and Paul Harvey, of Durham and Oxford respectively, read and patiently criticised later – 'penultimate' – versions. John Clegg, sometime curator of the White and Oates Museum at Selborne, answered questions on White as gardener. I could not have finished the book without the help of these specialists, although how I used their advice remains, of course, my own responsibility.

T. D. August 2006.

Foreign faunists are content if they can get a specimen,
& describe it exactly . . . but you will be able to shew
them in many instances that the life & manners of animals
are the best part of Nat: history.

Gilbert White

. . . the acutest observer by examining the dead body of
the water-ouzel would never have suspected its sub-aquatic
habits; yet this anomalous member of the strictly terrestrial
thrush family wholly subsists by diving – grasping the stones
with its feet and using its wings under water.

Charles Darwin

. . . anything that contributed to the flowering of natural history
is part of the history of evolutionary biology.

Ernst Mayr

Chapter 1

SELBORNE AND NATURAL HISTORY

1 A 'diversified' parish

White's respect for small and 'insignificant' creatures – the hirundines and harvest mice, the field crickets and young frogs he studied – was both genuine and partisan. According to those who knew him, he was a small, neat man. His allegiance to the parish he immortalised, the 'abrupt, uneven country, full of hills and woods and therefore full of birds'[1] was at one level an emotional allegiance. His description of his house and garden as a 'sheltered, unobserved retreat', in one of his rhetorical poems, is not to be taken quite literally; he regularly made trips and journeys and regularly put up friends and relatives, he was sociable by nature and enjoyed musical parties. But his satisfaction at getting back to his home territory after any considerable absence is plain. He would travel if there was good reason for it even in the depths of winter, but he seems to have felt entirely confident, or secure, only amidst 'Selbornian scenes'.

The sympathy and loyalty played their parts in his natural history career; they helped keep him steadily directed towards a particular, restricted subject matter; but they leave us unprepared for the originality, the independence of mind, which also typified him. He operated mainly within his own locality, and his career as a naturalist or committed naturalist began in the mid-1760s, but he was a methodical observer of the behaviour, as he often puts it, of the 'life and conversation', of animals and plants. On this evidence alone, he was out of step with his contemporaries:

> It is fair to say that all the eighteenth century naturalists, with the exception of Gilbert White, concentrated on descriptions and names. They were so keen to get the recognition of animals

> cut and dried that they nearly all put studies and accounts of the creatures in the field into second place.'[2]

But – and this will be one of the principal claims of this book – he was not merely an observer, his 'observing' was not a matter merely of watching living things and writing reports. Exploiting his special relationship with the parish – and in a sense proceeding in spite of it – he probed and questioned the 'lives' of his parish members and, beyond this, the organisation and working of the parish as a whole.

He was part of what can almost be called an eighteenth century British 'natural history' movement. Classification and anatomy dominated the higher reaches of this 'movement', and collecting, in one field or another, was the principal amateur naturalist's activity. White gathered wild and foreign plants into his garden, but otherwise made – or left – no collections. There are no rows of mounted beetles or butterflies, no cases of stuffed birds or trays of neatly labelled birds' eggs. During 1766, he catalogued the flowering plants of his district, covering more than four hundred species, but he left no books of pressed flowers. He was without the aquisitive and competitive – emulative – attitude of the true collector. He thought fossils should be given careful attention, and he kept various of the fossil specimens found by himself or passed on to him by his neighbours, but he made no attempt to amass 'ores and fossils'.[3]

He doubted the value of 'posting from place to place' in search of information. His view was that of a more recent behavioural worker: 'It is only by living with animals that one can attain a real understanding of their ways'.[4] Seasonal record-keeping – the recording in particular of 'first and last appearances' – was a feature of the popular nature study, and more often than not, this recording was a sort of collecting. Gilbert White was a committed recorder, but what recommended the Selborne district to him as a naturalist was its 'outdoor' diversity. The idea of an atomistic natural world was alien to him, but the parish, though it could be covered conveniently on foot or on horseback, included within it 'chalks, clays, sands, sheep-walks and downs, bogs, heaths, woodlands, and champaign fields'.[5] The weather of the locality varied; the resident and long-term observer could note what happened to given animals and plants, and what they did, in the range of seasonal conditions occurring during any year in Hampshire, and from time to time could add observations made during particularly warm summers or winters of 'rugged' snow, drought years (such as 1781) or years of persistent rain (such as 1782). Gilbert White's openness, his availability to other people and things, was a distinguishing quality, but in White the natural historian, an inquisitive scepticism complemented this openness. In the process of making discoveries he could doubt his own first-hand impressions; he could suspend judgement, in John Locke's phrase, until he had made appropriate comparisons.

Inevitably, and the advantages of his geographical situation notwithstanding, he met with obstacles as an observational naturalist in the eighteenth century. His elementary mistakes are instructive, two hundred years on. In 1767, for instance, he failed to identify a dead peregrine falcon which was brought him.[6] The incident is documented in the *Natural History of Selborne* – the bird was 'uncommon' in White's district but the carcass he examined was not seriously decomposed – although by the time the book was published he could also include a well informed report on the peregrine; 'candour' was indispensable to the true naturalist.[7] Again, he refers on a number of occasions to having seen 'yellow wagtails' in the Selborne parish in winter. This mistake, as we can assume it to have been, was never corrected.[8] Presumably, he confused the yellow with the somewhat similar grey wagtail, or grey wagtail in winter plumage; the grey is a resident in Britain while the yellow is, and no doubt was, a summer visitor only. Even more strikingly for the present-day observer, towards the end of his life, when he was gathering materials for a paper on the nightjar, he mistook the eggs of a nightjar for those of a snipe.[9] The shapes of the eggs alone should have put him on his guard, we tell him confidently; a snipe's eggs are sharply pointed, whereas those of the nightjar are flattened ovals, virtually without large and small ends.[10]

The explanation of mistakes such as these is, of course, that when White began his research the work of those concentrating on 'names and descriptions' was itself still at an early stage. Where would-be naturalists of our own time can take handy and accurate recognition guides almost for granted, the natural history books of White's day were bulky, unreliable and relatively expensive. They suffered from the illustrators' dependence on dead specimens and, if the illustrations were coloured, from a process of repeated copying from an original hand-coloured version.[11] White built up a small library from among the manuals then available, but for information useful to him as a field naturalist he was largely dependent on his own experience. Nightjars were seen and heard on Selborne Down in the late evenings in summer (the Down was and is partly covered with scrub and beech woods), and a nightjar often hunted for chafers around the oak tree in White's 'great meadow'; but he had not watched these 'wonderful and curious birds' *at the nest*, or rather, *at a nest containing eggs*.[12] The involved recognition problems with which he often had to deal appear, in the same way, from the error over the yellow or grey wagtails. He did not know the yellow wagtail was a summer visitor only (in the *Natural History* it is listed with his resident birds), and depending on age and seasonal plumage, the grey wagtail can somewhat resemble the yellow. Because of *these two causes combined*, he sometimes 'saw' yellow wagtails about the parish during the winter months.

Remarkably few errors can be found in his work, in fact; and he has an assured place in the history of zoology as someone who in several instances distinguished convincingly between creatures which, until that time, had been universally confused. He added the noctule bat and the harvest mouse to the list of British mammals, or was among the 'discoverers' of both. He was the first to methodically sort out the three 'leaf warblers' which were and are widespread in Britain in summer; as they would be called eventually, the chiff-chaff, the willow warbler and the wood warbler.[13] These small, olive-coloured birds are difficult to tell apart even with the aid of the binoculars which, in principle, give present-day observers an enormous advantage over White. They are difficult to tell apart, that is to say, even with the aid of binoculars, and even though – thanks to the work of pioneers – we can be prepared in advance for the minute differences between them.

The expertise White aimed at can still surprise us. The different songs or calls of the three leaf warblers first alerted him to their separateness. A 'good ornithologist', he says, should be able to distinguish his birds or bird species 'by their air as well as by their colours and shape, on the ground as well as on the wing, and in the bush as well as in the hand'.[14] He dwells on the gaits and flight-styles of a number of species and genera; but recognition was only an initial stage, for White, a means to other work. (If only for this reason, the principle 'shoot first and identify afterwards' was generally quite unacceptable to him.) The 'good ornithologist' will be interested in the feeding and courtship, the hibernation or migration, the attitudes to offspring and the social relationships of the creatures he deals with. In these more advanced enquiries too, exactitude will be essential. In one of the *Selborne* 'letters', he refers to the behaviour of a pair of barn owls which had nested and hatched young ones in the roof of Selborne church.

> About an hour before sunset (for then the mice begin to run) they sally forth in quest of prey, and hunt all round the hedges of meadows and small enclosures for them, which seem to be their only food. In this irregular country we can stand on an

> eminence and see them beat the fields over like a setting-dog, and often drop down in the grass or corn. I have minuted these birds with my watch for an hour together, and have found that they return to the nest, the one or the other of them, about once in five minutes.[15]

He seems to have watched birds and animals using only the naked eye, but from the quality of his detailed reports this inconvenienced him only rarely. In what was a secluded parish in his time, and in winter even an inaccessible one, he usually worked alone: at least – and though he could and did call on helpers from within the Selborne community – he was without a 'companion' in natural knowledge, an assistant who fully understood and shared his aims.[16] This did constitute a handicap, he readily admitted.

White was a gardener as well as a naturalist and a poet as well as an objective investigator; a romantic flourish was added to various of his activities. With his brothers, he built two 'hermitages' on Selborne Down, to be visited for tea parties and for the views they offered of the surrounding countryside, and he set up a large 'figure of Hercules' overlooking his outer garden. He copied the following lines from James Thomson onto the title pages of the early annual volumes of his *Naturalist's Journal*:

> I solitary court
> The inspiring breeze; and meditate the book
> Of NATURE ever open . . .[17]

Only, he recognised the limits of solitary research. A man trying to be sure of his facts but working exclusively from 'his own autopsia' can seldom make steady progress, he says, and in his own case a local colleague would have 'quickened his industry and sharpened his attention'.[18] The behavioural nature study he wanted to be part of would be both collaborative and organised. True advances would require a sharing and mutual monitoring on the part of widely – even internationally – scattered 'observers'.[19]

He never despaired of seeing at least the early stages of this co-operation. He tried to lead by example. He was a gifted letter-writer (that he may not have *liked* letter-writing has given some commentators a topic for discussion, but whether he liked it or not, he wrote a great many informative letters) and he helped other naturalists generously. Without being responsible for the general character of that work, he contributed much more to the later editions of Thomas Pennant's important *British Zoology* than Pennant (1726–98) admitted in print.[20] At the suggestion of the Hon. Daines Barrington (1727–1800), he submitted behavioural papers to the Royal Society, on the swallow, the house martin, the sand martin and the swift, which appeared in due course in the *Philosophical Transactions*.[21] If we include *The Garden Calendar*, he kept detailed fieldbooks for a period of more than forty years. For twenty-six of those years, from January 1768 to June 1793, he kept a *Naturalist's Journal* almost continuously, and where meteorological records were concerned, daily. (When he was away from home, Thomas Hoar, his loyal man of all work, 'kept an account of the rain'.) These diary/notebooks give us insights into his protracted work on particular natural history problems, and the *Naturalist's Journal*, at any rate, was intended for the use of others as well as himself. Finally, towards the end of his life, he published *The Natural History of Selborne* (1789), the collection of behavioural and ecological reports based on his letters to Pennant and Barrington. He had revised these reports at intervals, using his journal records, and had added other papers to them, notably on geology and meteorology.

He could send requests for information to acquaintances scattered over much of the southern half of Britain, and he encouraged his friends and relatives to make natural history observations and keep records, especially if they travelled abroad. His brother John was to be a considerable asset during his years at Gibraltar with the garrison. But he, Gilbert, remained always short of reliable co-workers. From time to time the 'help' he received proved a positive hindrance, as is clear from *Selborne* itself. Only White's side of the correspondence appears in those of the *Selborne* papers which are based on original letters, but Pennant and Barrington could both pass on untested, hear-say information. White was often informing them, in the event, but in doing this he had to proceed diplomatically. Barrington was a magistrate and a dilettante writer and naturalist where Pennant was an established zoological author, but both were fellows of the Royal Society. White and Barrington became friendly associates, but both of these, his principal correspondents, were of greater social standing than the Hampshire naturalist and curate.

On the positive side, he worked within an established methodological tradition (this will be considered in some detail in a later chapter); and though his remarks in the *Natural History* concerning his methods are sometimes misleading, we do not have to depend only on this source for our knowledge of these methods. For one thing, parts of the *Antiquities*, the lengthy appendix to *Selborne* as originally published, can still be read along with the more famous work. The publication of the papers now generally associated with White did not take place until the winter of 1778–9, because of his 'unassisted' thoroughness, and because he decided at a half way stage to include the results of his researches into the medieval manor and priory of Selborne. To attach these results to the *Selborne* 'letters' – to provide in *The Natural History and Antiquities of Selborne* what were in retrospect two books under one cover – was no doubt a mistake; but, from several viewpoints, medieval Selborne had a significant bearing on the eighteenth century parish. If antiquarian exercises were fashionable during White's adult years, his determination to accommodate this historical dimension also illustrates his own working attitude; information on other Selbornes helped him to look objectively at the Selborne of his time.

More obviously, the nature journals – these journals as White left them, including the meteorological records – augment and throw light on the *Selborne* 'letters'. Through the medium of the journals we can watch him as he gathers his outdoor materials; and often it was his journal-keeping that enabled him not only to respond directly to his surroundings but also to 'examine them comparatively'. With the help of his records, he extended and monitored himself, or compensated in some degree for the scarcity of reliable co-workers. The essential document is again the *Naturalist's Journal*, in its twenty-six – or twenty-five and a half – annual volumes. These journal/notebooks are free of blots and are particularly legible. White clearly took pleasure in his regular, methodical filing; and as the years followed each other – as questions were defined and as the parish and its weather varied, as the annual volumes accumulated and as the *Journal* itself suggested further 'puzzles' – he could make penetrating cross references from one of these sets of observations to another.

Comparative investigation would be among the possibilities enhanced as or when White became a collaborator and team-worker; but the use he made of records was exceptional in his day. His behavioural stance – that he concentrated almost exclusively on animal and plant behaviour – was exceptional, and was self-conscious, it can be emphasised. At least, he knew much more about the work of the other naturalists of his age than appears immediately from the work he published. (He had useful London connections, and, busy and alert, he kept himself informed. As we shall see, the experimentalist, physiologist and vivisectionist, Stephen Hales, was among the naturalists known to him personally.) He concentrated on the habits of 'summer' and 'winter' birds, frustrating though this could be, and on the minuscule life-styles of his crickets. He studied the behaviour of that 'bird of the twilight', the 'fern owl' or nightjar, and after dark he would visit his grass plots 'with a candle' to observe the activities of earthworms which had emerged – or as he points out, partially emerged – from the holes in which they were normally quite hidden from view.

Living things in general, we can almost say, interested him only in so far as their conduct could be examined. However, the notion of White as a *mixed* – or more satisfactorily, a *multi-levelled* – man, will appear repeatedly in the course of these chapters. He acknowledged but also shrewdly challenged various current ideas; he dealt judiciously with bad advice, but on occasion could reject good advice (as he did perhaps over the combining of the *Antiquities* with *Selborne*). He was sensitive and yet sociable, cautious and yet enthusiastic; he was considerate almost instinctively and yet, at another 'level', was insatiably 'inquisitive'. A tendency on the part of commentators to smooth out, or edit, this mixed man has been unfortunate, I shall insist; White's 'self-contradictory' character in part explains his natural history career. He hoped to be a team-member, and was always on the look-out for suitable colleagues. Such colleagues were rarely to be found, but the personal mixture included – along with a certain timidity – the moral pertinacity required if he was to fill out his scientific role.

Looking back on it, we can see his radical early gardening as a preliminary to his work as zoologist and botanist. Learning as he went, he developed a large, multi-purpose garden and smallholding despite what was a particularly intractable – chalk – soil. He had assistants, but in important respects he alone was responsible for this project; two of his brothers helped with the buying of land, but only after he had proved himself as a horticulturist. The ebullience of his 'early gardening' was not – and perhaps could hardly have been – repeated, and he fully recognised his vocation as a naturalist only in early middle age, but once he had recognised it, many of the habits of the gardener appeared also in the exponent of 'natural knowledge'.

2 The naturalist's apprenticeship

The *Garden Calendar*, the first of White's journals, is a record of the re-modelling and gradual enlarging of the Selborne 'garden'. White was born at Selborne, in 1720. He began his correspondence with Pennant and Barrington, and started keeping the *Naturalist's Journal,* only during the later 1760s, but he kept the *Garden Calendar* for part of each year from 1751 to 1767. After graduation and ordination, both at Oxford – and after various 'jauntings' about southern England – he assumed responsibility for the garden at The Wakes while still accepting temporary curacies. He had been awarded a college fellowship, but a later and perhaps half-hearted attempt to secure a provostship at Oxford was unsuccessful. He was settled at Selborne in the family house by the late 1750s. He served as curate at Faringdon, two miles to the south west of Selborne, for twenty-three years, but he could fulfil his duties there while living at home. From our own point of view, then, it is of relatively little importance that he became permanent curate at Selborne, his 'own' village and parish, only in 1784, nine years before his death.[22]

If conditions in the Selborne parish were conducive to White's progress as a naturalist, his gardening during the *Garden Calendar* years gave him his naturalist's apprenticeship. What strikes one first is the arduous character of many of the projects referred to in this early journal.[23] In 1751, the garden and orchard at Wakes, or The Wakes, covered perhaps two acres, with an additional garden area sketched out in one of the adjacent fields, but by the late 1760s White was responsible for a garden and 'farmery' covering between them perhaps eleven acres.[24] The total, only parts of which can be seen today, included a terraced garden with lawns and borders immediately in front of the house (facing Selborne Down and the Hanger) and a planted strip, the 'new garden', at the back of the house (against the street);[25] a shrubbery and a re-planted orchard; an 'outer garden' of perhaps six acres, with high, leafy hedges; a large meadow (the 'great meadow'), and two arable fields. The outer garden was traversed from west to east by a broad walkway and there was a pond in or near the outer garden. Two summer-houses or 'arbours' – not to be confused with the 'hermitages' on Selborne Down – overlooked the garden and fields, and for good measure, the 'figure of Hercules' stood on the far side of the property, below the Hanger. The Hercules had been made by John Carpenter the village carpenter, and painted to look like statuary, but taken with its pedestal, White tells us, it was 'more than twelve feet high'.[26]

Gilbert White had been co-creator with his brother John of the zig-zag path on Selborne Down, and he was eventually to add a one-storey parlour to the north west side of the house.[27] Later again, in 1780, he was to have the sloping 'bostal' path dug on the side of Selborne Down. During the *Garden Calendar* years, he finished off the terrace with a stoutly walled ha-ha, allowing the eye to range on unimpeded to the great meadow and the Hanger itself, and built a fruit wall at one end of the terrace.

He was to train vines on the walls of the house, to augment a vine on the fruit wall: he was a lover of trees and was especially proud of a 'great spreading oak', a large walnut, a Balm of Gilead fir and, in a small wood he owned on the other side of the village, a stand of 'fine beeches'. At one stage he tried his hand at grafting, and he replaced fruit trees and added new varieties. According to John Clegg, he grew over the years six different sorts of plum and seventeen varieties of pear.[28] In November 1759, he writes, 'Planted in the new Garden two standard Duke's Cherries; an espalier Orleans plumb; an espalier green-gage plumb; a dukes cherry against the north-west wall of the brewhouse; and a standard muscle plum in the orchard'.[29] These trees were put in the ground, he adds, the day they were taken from the nursery.

He mastered the cultivation of an enormous range of flowers and vegetables.[30] The pleasure he took in flower gardening, in particular, remained undiminished throughout his life, but during the first phase of his gardening, in October 1759, he could leave the following note concerning one day's work:

> Planted the irregular slip without the new wicket in the Garden with first two rows of Crocuss; a row of pinks; several sorts of roses; Persian Jasmine and yellow Do. several sorts of Asters, French-Willows; a curious sort of bloody wall-flowers; double Campanulas, white and blue; double Daisies; and a row against the hedge of good rooted Laurustines. Planted the back row of the part of the bank newly length'ned-out with blue and white Double Campanulas; and the border under the dining-room window with the bloody Double wall-flowers. Planted a bason in the field with French willows. Planted many dozens more of Coss Lettuce against the buttery wall, and down the wall against the yard.[30]

He perhaps had one or even two paid helpers on an occasion such as this; but the will and enthusiasm were his own.

He dug and prepared 'basons' where he needed a radically improved soil, and made extensive use of 'hand glasses' or circular cloches. The outer garden, divided into plots, was kept mainly for vegetable growing, although here too flowers had their places; a cheerful, elegant character pervaded even the more businesslike departments of his estate. Just as importantly, his hot-beds were made up in the outer garden each year, principally for the growing of melons and cucumbers. Again the scale of operations can be remarked on.

Typically, one of the 'fruiting', as opposed to propagating, hot-beds for his cucumbers was fourteen feet in length and 'two lights' wide, and he refers to making up a melon bed which was forty five feet in length. (Another year he made up a melon bed which was thirty six feet long and 'six and a half' feet wide.) The crucial ingredient in these hot-beds was fresh farmyard dung, and the amounts are carefully recorded. In the course of one season, in 1759, he used almost fifty cart-loads of fresh dung.[31]

Meteorological notes and summaries (though at this stage not daily records) appear in the gardening journal; but within the limits of his hot-beds and frames – his alternatives to heated greenhouses, which though they were quite outside his means were prominent features of the grander establishments of his time – he was replacing the weather of his region.[32] Notably, he was producing his own late winter and early spring weather. During much of the *Garden Calendar* era, he was still acquiring the skills needed in hot-bed cultivation; he documents his progress in detail as he tries out fresh techniques or uses the knowledge he has gained already in new ways. He started the hot-beds for his frame cucumbers at earlier and earlier dates, until he was sowing his first seeds, in pots set in the soil and 'tan' covering the dung in a nursery bed, at the beginning of January. The seedlings would be planted out in a previously prepared fruiting hot-bed in early February – often in a garden still covered with snow. Sometimes he would begin using one of the beds before the first violent heat had dissipated. He would then have to rescue the seedlings, and perhaps partly dismantle the bed to cool it down, or if the damage was severe, start his planting over again. But within a few years he was regularly cutting his first cucumbers by the middle of April; he was gathering cucumbers, this is to say, when a less ambitious gardener might still have been planting out his seedlings, even in hot-beds. His earliest date was April 5th, in 1763. He was due to travel to London at this time, to visit one of his brothers on business, and he carried this first cucumber with him. A week later thirteen more were sent up 'by the coach', and he proudly records the price – 'two shillings apiece' – cucumbers were fetching in London at this early date. During the previous winter in London, the young James Boswell had been paying a shilling for a beef dinner with beer, and at Selborne at this date two shillings might be a skilled workman's wages for a day.[33]

Early cucumbers and melons were achievements aimed at by professional gardeners in White's time. He corresponded with Philip Miller – the nurseryman and hot-bed expert – and visited his Chelsea garden.[34] White could bring melons to maturity at Selborne only in the early summer months, but he notes while still a tyro gardener: 'Received from Mr: Philip Miller of Chelsea about 80 Melon seeds 1754: immediately from Armenia; which he finds to be better than those that have first been brought to Cantaleupe, & thence to England'.[35] Cucumbers and melons remained prominent in the horticulture of his subsequent years, we can notice in passing, with quantities still being carefully recorded. Between April 18th and May 13th, 1787, he cut close on a hundred frame cucumbers.[36]

Frame cucumbers and melons require carefully managed temperatures (for the cucumbers, not more than 80°F and not less than 60°F); but White could give any of a range of other vegetables and flowers an early start, in cloches or small frames given over to the purpose or in the frames with the seedling cucumbers or melons. Radishes, celery, 'indian corn', and among flowers, hardy and half-hardy annuals might be helped forward in this way.[37] Where these special conditions could not be provided, and on behalf of his early fruit blossom and his laurustines and arbutuses, he found ways of fencing with snow and frost. He could derive benefits from a snowy winter, he learned. A settled covering of snow protected tender plants during a severe cold spell; it assisted the year-round gardener with, for one thing, his winter lettuces.

But large-scale gardening at The Wakes would have been problematical even without these winter programs. (White was made of sterner stuff than we have sometimes been led to expect.) Temperatures at Selborne, if they could vary considerably from year to year, were lower than present-day temperatures at the same place. They were lower than those at Bramshott, a few miles to the east of Selborne, or the 'villages about London', one of which White visited in the course of most years. He was unlucky with the weather; from 1764 to 1774 inclusive he had a succession of especially wet summers.[38] His soil, finally, was of a kind notorious among gardeners; it was a stiff chalk marl. Whatever may be suggested by the quantity and gratuitous range of his flowers, fruit and vegetables, he was working on ground some of which had never been gardened before, and with soil which had to be endlessly reconstructed for worthwhile gardening to be possible on it.

The Wakes lies on the south-west side of the village street. The ground on that side is 'good wheat land', White tells us in the first of the *Selborne* papers; but a gardening prejudice creeps in. While the soil on the *north-east* side of the street is 'a warm, forward, crumbling mould', that on the south-west side – so diversified is the parish – requires 'the labour of years to render it mellow'.[39] Chalk marl is fertile, but becomes sticky and heavy after rain. (It is wrongly, though understandably, referred to as a clay, in the paper in question.) If a warm, dry period follows the rain, it sets to a concrete pan, and then cracks – or as White says, 'chops' – badly. Concerning one of his fields, White reports in July 1765: 'Farmer Knight, having plowed Bakers Hill twice before, stirr'd it across today. The weeds are all kill'd, & the soil is baked as hard as a stone, & is as rough as the sea in a hard Gale: the Clods stand on end as high as one's knees'.[40] The knees, at any rate, of a small man.

Remarks on what had to be done with this soil continue periodically throughout the years covered by the *Naturalist's Journal* (1768 to 1793).[41] It was perhaps because of the difficulty of working it that the outer garden, with its walks and high hedges, was not maintained after the land passed out of the hands of the White family. Gilbert White kept large parts of his ground in the 'mellowed' condition. He made a study of manures and soil conditioners. At one time and another, he used rotted dung, tannery bark ('tan'), straw, spent hops, sand and cinders as ameliorators, and annually, at the beginning of each year, he sowed his meadows with huge quantities of wood and peat ashes. He made a special 'mould' for his hot-beds (i.e. for the top layer), including in the mixture a lighter soil from Dorton on the other side of the village.[42] He was instrumental in encouraging the greatly increased growing of potatoes in the Selborne region. He records in April 1765: 'Planted five rows of potatoes in Turner's garden, & put old-thatch in four of the trenches, & peat dust in one for experiment's sake. Exchanged roots with Mr. Etty, as his ground is so different: his sort came originally from me.'[43] Where the peat dust had been added, they got the better crop, we learn, in the autumn of the same year; and White was later to use peat dust – as it can be termed, a fibre and acid contributor – on his own ground annually with good results.

3 Wider intimations

Throughout White's adult lifetime his gardening effectively added to his income. Like his Oxford fellowship, it was among the factors enabling him to give up the serious consideration of paid employment at places remote from his home region. He could never become vicar of Selborne, because the Selborne living was administered by Magdalen College, Oxford, and he was a graduate and fellow of Oriel. The house in the village street came to him as a bequest, but he had little private money; Thomas White, the second of the White brothers, was the main recipient of the modest investments passed on by his parents, and notably his mother.[44] The skill Gilbert developed as a gardener and smallholder tipped the balance, we can say, enabling him to establish himself at Selborne in adulthood. With the varied garden produce added to his other resources, he could both live in the village and fulfil his responsibilities as a senior family member.

Some enlargement of the garden at The Wakes had been begun, albeit rather ineffectually, by White's father; and increased productivity on the part of gardeners and farmers was a self-interested but moral aim, by the second half of the eighteenth century. In the *Naturalist's Journal*, White quotes Virgil periodically as well as James Thomson, the Virgil of the *Georgics*, the Latin apotheosis of gardening and farming. The true horticulturist will be sensitive to the natural bias of his ground, Virgil had insisted:

> . . . plough not an unknown plain:
> First you must learn the winds and changeable ways of its weather,
> The land's peculiar cultivation and character,
> The different crops that different parts of it yield or yield not.
> A corn-crop here, grapes there will come to the happier issue:
> On another soil it is fruit trees, and grass of its own sweet will
> Grows green.[45]

White was influenced by this ideal, in his planting and harvesting and as he studied his local soils and weather cycles. Or rather, like many of his eighteenth-century English contemporaries, he was subscribing to and stretching the ideal. He studied his local conditions but worked indefatigably at re-modelling his soil; he took all manner of liberties with his crops and crop yields. Classical notions of gardening and farming influenced mid- and late eighteenth-century Britain, but the era was also well provided with 'moderns', or progressives, in both these fields. The enlightened horticulturist enhanced and corrected the features nature had provided, with the aid of recent science and technology.

The title *Garden Calendar*, or according to White's earlier spelling, *Kalendar*, is misleading. The book, made up originally of separate sheets stitched together, is a factual record such as would have been indispensable to someone attempting numerous gardening and 'farming' projects at the same time. (It has little or nothing to do with the calendar-compiling White did take part in during 1776, which showed the 'usual' sequence of botanical events in a named district.[46]) But humanity keeps breaking through in the course of this first journal. White wrote laudatory poems about his village, and was proud of his fortified blackcurrant wine. He adds further stylish features to his garden and grounds, some 'oil jars' and at one stage a 'Chinese fence'. He presides over 'melon feasts', as he calls them; towards the end of 1763 'Drank tea twenty of us at the Hermitage: the Misses Batties and the Mulso family contributed much to our pleasure by their singing and being dressed as shepherds and shepherdesses.'[47] Even in his notes on the routine work, if gardening was ever routine to White, an intense satisfaction shows through his 'factual record'. He multiplied the challenges he had to deal with as a gardener at The Wakes; at this distance, his almost headlong 'early gardening' cannot be wholly explained; but during this sometimes unsettled period of his life, clearly, he immersed himself in gardening when he could.

He was already an observer of a sort, during these years, and the *Garden Calendar* is not confined only to gardening topics; entries on house martins, field crickets and wild as well as garden flowers appear towards the end of it. More generally, and though his eventual career could not have been forecast during most of this period, his subsequent views as naturalist – and even animal psychologist – can be found in embryo, in various of these 'garden' notes and comments. Exploiting the dependence of any plant on a wide range of external circumstances, and whether or not he thought about it, he experienced at first hand the relative instability of individuals and even species. He from time to time suffered the complete failure of a garden crop. On one occasion he started his frame cucumber seedlings in mid-December and, for all his care and persistence, was unable to get them past the transplanting stage. He was to imply eventually that animals are consequences of what they bring with them into the world and of the conditions they then experience, where what they do may alter or add to these conditions; but he appreciated that animals, as much as plants, may often not survive an abrupt or extreme change of conditions.

An incipient literary style, and the exact detail we come to associate with Gilbert White, are apparent in these early gardening, smallholding and domestic records. (For a time he kept note of the ingredients and quantities used in his regular wine- and beer-making in the *Garden Calendar.)* He sometimes challenges the 'obvious' almost indiscriminately, during this early period, but the entries were such that he could also make a genuinely investigative use of the results of the successive years. (When, in 1767, Daines Barrington sent him the printed diary blank which was to become the basis of his *Naturalist's Journal*, he was already an experienced record-keeper and record-user.) The potato trial, the trial with the thatch and peat dust added as ameliorators, mentioned above, was more than a casual 'experiment'; and a modification he introduced into one of his forcing frames or, properly, his manner of testing this modification, would today be accepted as rigorously experimental. Using hot-beds and frames early in the year – and with his melons, even during the 'warmer' months – he had a problem with ventilation. The working dung produced fumes which seemed detrimental to the young plants, but the temperature outside the frames might be so low, compared with what was required inside, that if the lights were left open for any length of time even at mid-day the necessary warmth inside could not be maintained. (At night, far from being left open, the lights had to be heavily covered over with mats or sacks.) To let these fumes out, White fitted a tin chimney to one of his 'nursery' frames, and in February 1758 he added a pipe at the bottom of the frame through which clean air could be drawn in, a pipe passing first through its own dung-filled heating box. Several weeks later he tested the air in this and another of his frames after they had both been closed and covered up for some hours, using a lighted candle. He records:

> On putting the candle down a few Inches into that frame which had leaded lights & no chimney, the flame was extinguished at once three several times by the foul vapour: while the frame with the tiled lights & chimney was so free from vapour that it had no sensible effect on the flame. I then applied the candle to the top of the chimney, from which issued so much steam as to affect the flame, tho' not put it out. Hence it is apparent that this Invention must be a benefit to plants in Hot-beds by preventing them from being stewed in the night time in the exhalations that arise from the dung, & their own leaves. The melons confirm the matter, being unusually green and vigorous for their age.[48]

His generally 'inquiring' attitude is illustrated in this minor episode, but what should be noticed too is his use of the other hot-bed and frame – that to which no ventilator had been fitted – in checking the ventilator. The two hot-beds and frames were more or less comparable, we can assume, except that a ventilator had been fitted to one of them; and therefore – when he had carried out tests on both – he could claim unequivocally that the improved quality of the air in the frame with the ventilator *was due to the presence* of the ventilator. The improvement had come about, but in addition he had used a 'control', as it would now be called, in investigating the improvement. He could have pointed to interesting evidence without the use of the control, but because a control was brought into the inquiry, he could point to essential, conclusive evidence.

'Controls' are indispensable to rigorous investigation, and the use of these safeguards was typical of the later – zoological and botanical – White, it can be shown. No theoretical discussion of controls is to be found in his writings, but they became a standard feature of his work on wild animals and plants in their usual habitats. In his later activities enthusiasm is present but is strictly managed. The eventual naturalist uses controls as he probes deeper into 'manners and modes of life' – and as he throws suspicion on his own loyalties and preferences.

Chapter 2

NATURE AS AN 'ECONOMY'

1 The parish community

White was the eldest of seven children who lived on into adulthood, five brothers and two sisters. Of these only he made Selborne his permanent home.[1] Thomas and Benjamin became London businessmen, though by the mid-1770s Thomas was a retired businessman, and both married and brought up families at South Lambeth. Gilbert – the country curate and bachelor – stayed with one or other of these brothers when making his London visits. He first met Thomas Pennant through Benjamin, and it was Benjamin White, a bookseller and publisher, who in due course brought out *The Natural History of Selborne* (1789). John and Henry, the younger brothers, both became parsons, John taking up his Gibraltar post and, after some years abroad, and encouraged by Gilbert, developing ambitions as a naturalist parson. John's *Natural History of Gibraltar*, though it seems to have been completed, was never published, but we have a selection of the letters written by Gilbert to his brother regarding this Gibraltar project.[2] The work of advising John White exercised and extended Gilbert, and notes on migratory birds, in particular, provided by John were to be of great use to Gilbert. The brothers all took some interest in natural history, in keeping with the times, and so did both of White's brothers-in-law; Thomas Barker, who married Anne White, was a noted meteorologist. Like Gilbert himself, they continued to think of The Wakes as family property even after he had become its proprietor. Gilbert ensured that it remained a family gathering place, and Thomas and Benjamin – though perhaps somewhat patronising in their attitudes to the oldest brother – helped him to secure the pieces of adjacent land.

Clearly, the parish itself provided White with a family. Just as importantly for our own enquiries, the Selborne parish was still a locality in which the human ecology was continuous with the non-human; they reflected and depended on the same resources. Selborne was and is situated 'half way between Alton and Petersfield', small country towns themselves lying some eight miles apart. Though each of these towns adjoined an important coach road in White's day, Selborne was served only by cart-ways and narrow lanes; a friend from the naturalist's college days, John Mulso, would ask to be met by a guide when he came to visit him, for fear of getting lost. But a carpenter, two grocers, a shoemaker and saddler, a butcher (in front of whose yard White planted some lime trees) and two blacksmiths all had shops in or near the village street. According to White, 'We abound with poor, many of whom are sober and industrious, and live comfortably in good stone or brick cottages, which are glazed, and have chambers above stairs: mud buildings we have none.'[3]

The relative prosperity of the parish derived principally from its agriculture, or the mixed, alternating character of its agriculture. For reasons which will appear later on, there was nothing in the parish comparable to the 'great farm' at Newton Valence, a neighbouring village, where more than a hundred acres of barley alone might be waiting for the harvesters,[4] but White could refer knowledgeably to Grange Farm, Priory Farm, Norton, Burhunt or Temple Farm, properties of perhaps sixty acres on average. Within the parish boundaries there were pasture and arable lands, 'sheep downs', hop-gardens and deciduous woods. A range of specialised occupations went with this varied land use, but farming at Selborne also presupposed a numerous, adaptable work force. Much of the work was seasonal; during hop-picking, to take the obvious example, large numbers of workers were needed for only a few weeks; and some could be part-time. Married women and periodically children could add to a family's income. Most significantly, because of this varied husbandry, the labouring men of the parish, moving when necessary from one sort of work to another, could be fully employed for all or most of the year.[5]

If only because of what has been called his 'genius for intimacy', White was a party to the finely drawn inter-connections obtaining in the parish and parish community. As a naturalist or in several of his roles at once, he was out and about in his locality almost daily; we can find him in attentive conversation with well-diggers, ploughmen, shepherds or village housewives. Some of the parish residents must have known him since his childhood, but he seems to have easily gained the trust even of chance acquaintances.[6] When a carriage, or carriage horse, broke down as he was returning cross-country from

Andover he waited for a replacement in the home of a carter and his wife. He was soon eliciting their recollections of the great bustards which in earlier years had appeared in parties on their local downlands and an account of the – less welcome – 'Norway rats' which had recently invaded their buildings.[7]

The diversity of his knowledge of country matters was due in large part to these personal contacts. He could provide a detailed description of the domestic manufacture of rush lights, the labourer's alternative to candles, which burnt more or less efficiently depending on how they were made.[8] He notes that the barking of felled timber is easier in the spring than at other seasons, and relates this to his own theory and practice as a gardener.[9] Local people brought him wildlife reports and living or dead specimens; he treated their reports with caution, but his hypotheses as a naturalist often reflected the contributions of these willing informants. A peat-cutter, a woman who lived 'just at the foot of the Hanger' and a farmer were involved in the mistake, and then the correction of the mistake, concerning the eggs of the nightjar and the snipe, and popular prejudices against the nightjar (the belief, for instance, that it was 'injurious to weanling calves'), were one reason for his having started his investigation of this 'wonderful and curious' bird.[10]

For much of the time he had only one full-time garden assistant, Thomas Hoar, but when he needed it, he could easily get extra help.[11] He largely provided for his own household, and fed the visiting friends and relatives, from his garden and smallholding,[12] but there must have been a surplus, considerable though his own total requirements were. This surplus was not marketed in any formal manner, but he took part in what was in effect a continuous local bartering of farm and garden produce.[13] He encouraged co-operation; an example appears in the way he procured the large quantities of dung he used each year in his hot-beds. He could provide some of this dung himself, but most of it came from neighbouring farms, and during the *Garden Calendar* years he kept precise records of the numbers of cart-loads of dung delivered, where they had come from, and when they were used. Oddly as it seems at first, he describes these additional supplies as 'borrowed'. But in the fresh condition in which the dung was invaluable to the hot-bed specialist it was unusable by the farmers, and when the hot-bed man had finished with it, and when it had been heaped and allowed to 'rot down', it was in the right state for digging or ploughing into the ground as a general manure. White made a payment to the farmers but this was for carting and a temporary, restricted use of the dung, we find. In due course, usually during the next January or February and in the carts which had again been delivering fresh dung, much of it was returned to the farmers who had first supplied it. They then used it on their fields, we can be sure, even if by this time they were a little confused.[14]

According to White's notes, 'Saxon dialect' was distinguishable within the speech of the Selborne district.[15] The effects of ancient dispensations were still apparent in the organisation of the parish and parish community. The 'poor' maintained their traditional fuel and grazing rights and the right to let their pigs into the woods in autumn, and we read of gleaning and the regular gathering and selling of acorns. Those of the villagers who were 'copy-holders' had the use of the large common fields, the King's Field, the North Field, the Fore Down and the Nine Acres. They strip-cultivated these, and after harvest, by annual arrangement among themselves, 'threw them open' to cows and geese. (These common fields were still intact at the time of White's death, and remained uninclosed until well into the next century.[16]) Alternatively, a copy-holder might lease his right to the use of the same land to one of his neighbours, and as a consequence, Selborne could boast a number of smallholders, or 'little farmers', as well as the more obviously established farmers.

The local lanes related to the contours of the parish landscape, and the agriculture corresponded closely to the varied geology described in the early *Selborne* papers. The disintegrated freestone or 'white land' produced the finest hops, while the gault, the 'strong land', if it was appropriately limed, was the most suitable for wheat. The local weather, seasonal or unseasonal, directly affected the parishioners whatever their stations; drought was inevitably a cause of anxiety. Wells and well-digging feature recurrently in White's journals; the spring supplying the main Selborne stream 'never failed', but in January 1789, during a period of both drought and hard frost, White's near neighbour Timothy Turner deepened his well by nine feet.[17] Though most even of the labouring people had more than one means of support, the effects of persistent rain were matters of similar concern. In the wet autumn of 1792, as well as those trying to harvest the hops and corn, the naturalist records,

> the peat-cutters are great sufferers: for they have not disposed of half the peat & turf which they have prepared; and the poor have lost their season for laying in their forest-fuel. The brick-burner can get no dry heath to burn his lime & bricks: nor can I house my cleft wood, which lies drenched in wet. The brick-burner could never get his last makings of tiles & bricks dry enough

> for burning the autumn thro'; so they must be destroyed, & worked up again.[18]

Methods of guarding against tainted meat in summer – White's answer was a 'pendant pantry' hoisted into a tree to catch the breeze – and frozen bread, meat, cheese and potatoes in winter, are referred to in the journals. Selborne was healthy, the air was 'free from agues', but living in the parish 'comfortably' throughout the year, in the second half of the eighteenth century, required an exacting range of skills and experience.

2 Interdependent creatures

Details such as these are recorded in the successive volumes of the *Naturalist's Journal*, the volumes I shall refer to also as White's annual portraits of Selborne. The 'diversified' and yet loosely organic village and parish, where it was not the subject of his work, was the constant background to it. It was a received opinion in his day that nature at large was a kind of organisation, a combination of mutually related parts, and he was aware of many examples among his plant and animal subjects of interaction and interdependence. Diverse and seemingly quite unconnected living things could influence each other through hidden intermediaries; and if plant and animal populations were affected by soils and weather, soils and even weather might be affected by these populations.

A grazing succession could be observed among various farm animals. Sheep could eat well in a pasture cows or horses had finished with, and geese could feed on grass which has been close-cropped by sheep. In the way fish are helped to survive in some ponds on the edge of Woolmer Forest, we can see how 'Nature, who is a great economist, converts the recreation of one animal to the support of another'. The ponds are sandy-bottomed and 'sterile', but the fish, White believes, are indebted to the numbers of cattle which stand in the water in warm weather to cool themselves. The cattle drop their dung in the water and the insects which nestle in this dung constitute an important food supply for the fish.[19] While pursuing their own ends, the most 'incongruous' creatures may benefit each other. As he records during one August:

> While the cows are feeding in moist low pastures, broods of wagtails, white & grey, run round them close up to their noses, & under their very bellies, availing themselves of the flies, &

> insects that settle on their legs, & probably finding worms & larvae that are roused by the trampling of their feet.[20]

With the aid of this brief description, we can almost see the ponderous cows and the tiny, running birds.

In a celebrated paper, White insists on the value of the common earthworm, his example *par excellence* of a creature which, 'though in appearance a small and despicable link in the chain of Nature, yet, if lost, would make a lamentable chasm'. Worms are a major food source for 'half the birds and some quadrupeds', and they do little of the damage attributed to them by gardeners and farmers. (The culprits are really slugs and leather jackets.) They turn the soil, moreover, and by their ceaseless boring contribute greatly to its drainage and ventilation. They draw leaves and grass into their holes, and having eaten this vegetable material, they manure the soil. Humble but innumerable, worms are 'the great promoters of vegetation, which would proceed but lamely without them'.[21] They more than earn their living, as it has more recently been expressed, in elaborate 'economies'.

White's paper on the earthworm pre-dated Darwin's book on the same subject by more than a hundred years.[22] But White notes too that cereal plants have played a vital part in human history, and that foliage itself, especially the foliage of trees and grasses, is indispensable to certain natural economies. He makes no attempt to explain the immediate function of foliage, and he is not thinking here of its importance as compost or as an animal food. Rather, foliage effectively mulches the ground, and frequently acts, in addition, as an 'alembic', condensing out the moisture carried in water vapour.[23] At Selborne, as in some other parts of the world, foliage materially affects the water supplies of the locality.

Very long periods of drought, which happened only rarely in White's region but which he annotated with particular care, provided him with much of his evidence concerning this condensation. One such period occurred in the summer of 1781, and followed upon a partial drought during 1780. Many of the wells in the parish failed during this summer, and the eggs of butterflies 'dried and shrivelled to nothing'. The reapers were halted by only one or two brief showers during the entire course of the harvest, and in October, 'the ponds in the vales are now dry a third time'.[24] During the autumn of this year the (water) mills were often unable to grind, so that as one result there was a shortage of the barley meal normally used by the villagers to fatten their winter hogs, the hogs they killed in October or November, this is to say, to be tubbed and salted for use during the winter.[25]

Some of the damage – for instance that to White's peach and nectarine trees – was permanent, but Selborne survived this drought better than some of the surrounding parishes, and better than a stranger to the district would have expected. If most of the crops were adversely affected, the crop of acorns that year was heavy; the poor were able to 'bring forward' their hogs, therefore, in spite of the shortage of barley meal.[26] The Wellhead spring, the main source of the Selborne stream, remained as reliable as ever throughout 1781; we read of farmers sending waggons two or three miles to the Selborne stream for water.[27] Moreover, heavy dews and what White calls 'dripping seasons', which could occur in the parish even during periods of drought, continued off and on during this arid summer.

'Swimming fogs' might gather over the open parts of the Selborne parish during the night, and if this happened, and even though no rain had fallen, the field crops and isolated trees might soon be literally running with water.[28] 'Dripping' could interrupt hay-making and could badly blacken hay which had been cut but not carried, but that it also saved crops which would otherwise have failed is clear from a number of *Journal* entries: thus,

> No one that has not attended to such matters, & taken down remarks, can be aware of how much ten days dripping weather will influence the growth of grass or corn after a severe dry season. This present summer 1776 yields a remarkable instance: for 'till the 30th of May the fields were burnt-up & naked, & the barley not half out of the ground; but now, June 10th there is an agreeable prospect of plenty'.[29]

During 1781, the importance of condensed water vapour appeared if only from the fact that the 'little upland ponds' – by contrast with the lowland ponds – remained a viable water supply: as White was to work out, condensation was the means by which these upland ponds were replenished. But for the present, grasses and deciduous trees demonstrate this alembic effect most clearly. The role of trees as condensers appeared from White's own observations and records (this although, as he says, trees also 'imbibe' great quantities of water); and it was implied too in the reports of North American colonists. They had grubbed out deciduous trees in vast numbers in establishing their farms and settlements, and in several instances, following this deforesting, he understood, the rivers which had once driven their corn mills were reduced to the merest trickles.[30]

The extent to which events in the lives of animals and plants coincided with seasonal changes, was of constant interest to him.

'This evening chafers begin to fly in great abundance', he notes in May 1782; 'They suit their appearance to the coming-out of the young foliage, which in kindly seasons would have been much earlier.'[31] Echoes of John Ray (1627-1705), the English taxonomist and 'universal naturalist', occur repeatedly in White's note-taking. The 'excellent Mr Ray', the father of true nature study in Britain, as the later naturalist also calls him, had emphasised the adaptedness of many creatures to the environments they are normally found in. Thinking about the 'hirundines', White remarks that their flight styles vary from one species to another, the high, speedy flight of the swift (which he includes within this genus, though with careful reservations), contrasting remarkably with the vacillating, butterfly-like flight of the sand martin. Perhaps each of these species feeds largely on its own kinds of flying insects, he suggests, and perhaps the flight styles of the birds are 'influenced by, and adapted to' the behaviour of these insects.[32]

White can refer to 'chains' of natural events, with the seeming implication that natural economies are mechanical arrangements of some sort; but in practice he also rejects simple causal concatenations. In 'economies' as he describes them, an effect may be traceable to *numerous* causes and a cause may give rise to *numerous* effects. Although fluctuation in food supplies must be a major factor in bird migration, 'food alone' does not account for bird migration, he decides.[33] Wasps, while they abound 'only in hot summers', do not abound in all hot summers; something in addition to heat must be requisite to their emergence or survival.[34] (Although wheat, and in another note a grapevine, requires sunshine or warmth, if it is to crop heavily, sunshine alone will not ensure a good harvest.[35]) The commencement of hibernation in a hibernating species usually correlates closely with a given temperature, he finds, but again, one correlation may not make an explanation. In the case of slugs, anyway, those in his cellar become torpid coincidentally with those in the garden outside; 'hence it is plain that a defect of warmth alone is not the only cause that influences their retreat.'[36]

White was familiar with John Ray's *Wisdom of God in the Creation* (1691),[37] and while he was still concentrating principally on gardening, he had bought Joseph Butler's *Sermons* (1726) and the same author's *Analogy of Religion* (1736). The world and 'the whole natural government of it', Butler had written, seems to be 'a scheme, system, or constitution, whose parts correspond to each other, and to a whole; as really as any work of art, or as any particular model of a civil constitution and government.'[38] Neither Butler nor Ray described nature as all harmony, and conflict was one feature of the living world as White encountered it. (Far more young birds were reared in his neighbourhood each year than could possibly find food there as adults, for example, and a certain 'jealousy' therefore ensued.) But a functionality or 'purpose' appeared in the way the parts of a 'natural economy' related to each other. No part acted quite independently of at least some of the other parts, and when one part was changed, the whole might sooner or later be affected. When some beeches were cut down on Selborne Down, wild strawberry plants sprang up all over the newly exposed area. The seeds, White implies, which had perhaps been lying in the ground for years, had been ready but had had to wait for the opportunity provided by the increased light and warmth.[39]

In important respects, White himself belonged to the parish he was determined to examine; and if a parish such as his own was a complex but organised 'system', so was the 'life and conversation' of an animal or plant. How he nonetheless pursued his enquiries appears from the *Naturalist's Journal*, in particular; to an important extent, the articulated operation of his subjects dictated his investigative methods. White was without philosophical ambitions, but while doing his own work he was also conducting a dialogue of a kind with John Ray. In an always respectful manner, he was both acknowledging and diverging from the opinions of this parent figure.

Chapter 3

THE 'OUTDOOR' METHOD

1 Observations and investigations

From time to time, White seems to contradict himself in *The Natural History of Selborne*. Though the first letters to Pennant and Barrington were written in 1767 and 1769 respectively, and though White edited and added to these originals in incorporating them into the eventual *Selborne* (they became the principal dated letters of the text), he did not altogether re-write them for publication. What is asserted in one 'letter' may be qualified in another, and varying suggestions may appear in the course of the same letter.

Intentionally or otherwise, White leaves his readers work to do. Sometimes we have his answers to questions raised by the other two naturalists, but have to work out for ourselves what the questions were. He hardly refers explicitly to Barrington's sweeping but ill-informed contributions to the hibernation/migration debate, but in several of his 'letters to Barrington' he is responding directly to this correspondent's opinions on the withdrawal of the 'summer birds'.[1] White's remarks concerning his own methodology were often responses to the attitudes of other naturalists. These remarks should be quoted with some caution, therefore, although even taken at their face value, they give little encouragement to a 'merely passive' observing. White works largely within what he calls the narrow sphere of his own parish, but in an age with which Linnaeus is easily associated, an age concerned more with collecting and classifying than with watching live creatures, he insists that,

> Faunists . . . are too apt to acquiesce in bare descriptions, and a few synonyms: the reason is plain; because all that may be done at home in a man's study, but the investigation of the life and conversation of animals, is a concern of much more trouble

> and difficulty, and is not to be attained but by the active and inquisitive, and by those that reside much in the country.[2]

An established countryman, White agrees with J. A. Scopoli, the naturalist of Carniola, that,

> as no man alone can investigate all the works of nature . . . partial writers may, each in their department, be more accurate in their discoveries, and freer from errors, than more general writers; and so by degrees may pave the way to an universal correct natural history.[3]

By 'partial' naturalists he means regional naturalists first of all, and when speaking from his own first-hand experience he does not indulge in false modesty. He could contradict current authorities. Scopoli himself was not so 'circumstantial and attentive' regarding habits as White could have wished; house martins do feed their fledged young ones while in flight, for example, although they achieve this by 'so quick and almost imperceptible a slight' that only a persistent observer will have witnessed it.[4] Emphatically, and despite traditions Linnaeus apparently accepted, the cuckoo is not a bird of prey, and the smaller birds are not safe from hawks 'while the cuckoo sings'. To appreciate this last truth one has only to walk in the country, White implies: during the early summer months as much as in any others, the tell-tale feathers of small birds can easily be found 'in lanes and under hedges'.[5] A natural history writer should pass on nothing from mere popular report, he adds, without also expressing 'a proper degree of doubt'.[6]

White wanted observational data which had been personally ascertained by the contributor. His own observational skills were exceptional, whether or not he sometimes used a telescope.[7] In May, as swifts are sailing 'at a great height from the ground', it is possible to watch these small birds treading or copulating on the wing, he reports.[8] (He was the first to record this aerial copulating, but that it does take place has been amply substantiated since his time – by watchers using powerful glasses.) In his notes on the barn owls which returned to their young in the church roof at five minute intervals, the fact is included that they pause before entering the nesting place and shift the prey they are carrying from their claws to their beak. They then use their claws, he says, 'to take hold of the plate on the wall as they are rising under the eaves'.[9]

He was an unhurried observer – 'hurry of spirits' was condemned by White the natural historian, despite the attitude which had sometimes governed White the early gardener – where this might be quite the opposite of a lazy one. He applauded foreign travel, at least for others, and was an avid reader of travel books, but naturalists should not depend on flying visits to the regions they then wrote about. (John Ray and his young patron and assistant Francis Willughby had been enthusiastic travellers, but White was thinking here of his own touring contemporaries.) He promised to send Daines Barrington details of the periods during which each of thirty bird species sang or could be heard in the Selborne district, but when he began putting the facts on paper he found he was insufficiently familiar with some of the singers concerned. He carried a list of the birds to be covered in his pocket for many months, therefore, and 'as I rode or walked about my business, I noted each day the continuance or omission of each bird's song'.[10]

The precise chronological schedules which appear in the first two 'letters to Barrington', relating to the ornithology of the Selborne parish, provide answers to particular questions. There is no general theoretical scheme to which findings have to be adjusted, in *The Natural History of Selborne*; as Walter Johnson says of White, 'No man was less under the sway of ruling prejudices'.[11] But while White responded to his surroundings, he also returned repeatedly to selected topics; his writings, including the *Selborne* papers, are fragmentary in that they are discontinuous, but various areas of puzzlement – vegetation and condensation, the withdrawal of the hirundines, 'congenerous' species in diverse habitats and the part played by learning in behavioural adaptation were among the most important of them – are especially explored. In the *Natural History,* his extended work on a specific problem can be followed in his account of the *ring ouzels*. He recorded these black and white members of the thrush family first merely as rare, occasional birds in his region, but once he was on the watch for them he found that they made brief but reliable bi-annual appearances. They could be seen in the parish for a day or two in April, and were present again, usually for a rather longer stay, in early or mid-October.

Making sense of these periodic sightings, and eliminating various possible but false explanations, took several years, even though he had in fact suggested the correct solution at an early stage. The handbooks of the time made no reference to seasonal movements among ring ouzels, but White suspected he was dealing here with a 'new migration'. When he put it to Thomas Pennant, however, that the birds occurring at Selborne were travelling to and from more northerly parts of the country, Pennant provided 'evidence' rendering this explanation unlikely. Pennant lived in Wales, and he reported it as a fact that ring ouzels both nested in Wales and remained there throughout the winter.[12] The birds White observed appeared twice in the course of each year; if they were not summer migrants travelling to and from more northerly sites, perhaps they were *winter* migrants, he suggested next, moving *westwards* in autumn from the colder parts of Europe and returning *eastwards* in spring.[13] He then discovered that ring ouzels nested on Dartmoor, to the west of Selborne, and were not seen there during the winter months.[14] He tentatively returned to his first proposal; Pennant's 'information' had been quite untested, he came to realise. The birds made their autumn appearance at Selborne and at various points along the Sussex Downs; and another correspondent, an 'intelligent person', confirmed that they nested in the Derbyshire Peak District, to White's north, and left that locality in the autumn.[15] By October 1770, and as ring ouzels again showed themselves at Selborne, White could declare: 'Let them come from whence they will, it looks very suspicious that they are cantoned along the coast in order to pass the channel when severe weather advances. They visit us again in April, as it should seem, in their return.' On this occasion, though, he was writing to Daines Barrington.[16]

By the same methodical eliminating process he worked out that the 'rock pigeon' (rock dove) was the wild forebear of the domestic dove, and made perceptive suggestions as to the functions of winter flocking in finches.[17] He 'loved to trace' natural appearances, but he would allow himself a positive conclusion only if his search had been exhaustive. What I called earlier his characteristic appeal to controls belongs with this systematic testing of possible answers. He was pleased to hear that Pennant had spent 'some considerable time' in Scotland on a zoological visit, and he thought some young men of fortune should spend a year in Spain investigating its natural history, but *mere time*, he knew, would not necessarily produce conclusive results.

Patience was a virtue, but thoroughness was a greater virtue. White was 'comparative' as well as observational, I have insisted. Keeping to *Selborne* for the moment, an example of his 'use of a control' can be found in the paper in which he responds to a claim concerning the cuckoo accepted by the anatomist F. A. Herissant.[18] The cuckoo leaves its eggs with other birds – or is the exception to the incubating behaviour among birds which 'otherwise seems invariable' – because it is physically debarred from incubating, Herissant had said. It does not and cannot sit on eggs and hatch them, because its crop, instead of lying above or in front of the sternum, is placed behind it, over the bowels and making 'a large protuberance in the belly'. White doubted the validity of this explanation intuitively, but he did not reply merely on the strength of this intuition. He too dissected a cuckoo, and found Herissant's physiology to be more or less correct. But, he adds, 'the test will be to examine whether birds that are actually known to sit for certain are not formed in a similar manner'. He next dissected a nightjar and a hen harrier; both were 'known to sit for certain' and each, for good measure, had several physical features in common with the cuckoo. He was conducting *a* test here, he might have said; if on examination these other species had proved *not* to be like the cuckoo in the matter of the crop, sternum and so on, then their incubating behaviour would have had little bearing on Herissant's thesis. But – as he thought they might be – they were the same as the cuckoo in these respects; and Herissant's case 'falls to the ground', therefore. The three species were alike as regards these internal arrangements, and that the nightjar and the hen harrier did incubate their eggs constituted the control. The 'protuberant belly' and so on could not be said to render incubation impossible, and for all White knew, might not even inconvenience the two incubating species.

'Enquiry' is raised to the level of rigorous enquiry here, but just where does the rigour appear? Despite the use of the nightjar and the hawk, the investigation did not turn on 'analogous' argument, firstly. White was fully aware of the dangers of depending on analogy in natural history and himself warns us against them; analogy, though it can suggest hypotheses usefully, can provide no conclusive evidence. (He was using the nightjar and the hawk in connection with the claim that a given anatomical arrangement rendered *a bird* incapable of incubating.) But secondly, although the investigation involved physical dissection, it was not physical dissection which made it rigorously scientific.

The parish naturalist dissected more often than we might expect, but in offering an alternative to the prevailing nature study of his time he opposed not only collecting and classifying for their own sakes but mere anatomical research; even live birds in captivity might be irrelevant to the 'natural knowledge' he wanted to encourage. He occasionally obtained specimens for the purpose of dissecting them, and he frequently dissected dead birds or animals which he had found or which were passed on to him. Generally, he did so as a way of finding out what these animals had recently eaten – or whether they had recently eaten: swallows and martins perhaps hibernated, and they were said to be emaciated when they first appeared in spring, but those he opened and examined on at least two occasions were 'fleshy and fat'. He dissected, then, when doing so might throw light directly on behaviour; in general, anatomy was useful only if it facilitated the study of live performance. (Even his brother John, attempting his survey of the fauna of Gibraltar, who was bound to collect and sometimes dissect, was told in addition; 'learn as much as possible the manners of animals'.[19])

Both Herissant and White could have brought colleagues into their examinations; perhaps both did so. But White's enquiry was 'rigorous' principally because of what he did and Herissant omitted to do. He made use of controls, here safeguards against mere coincidence. His enquiry was conclusive, if conclusive in proving a negative. The attitude this controlling expressed was scientific, but – rather than confining the researcher to captives or cadavers – it was quite compatible with an attention to 'outdoor' habits. White lived among the animals and plants he studied, and was interested in what he could learn only as someone who lived among them; but his 'learning' included recording, and an investigative use of records.

Sometimes the parish would present the resident naturalist, anyway, with ready-made control situations. In June 1784, and in a case covered fully only in the *Journal*, a violent hail storm struck the north-western part of the parish, but reached no further than the centre of the village. The hail badly damaged the wheat-fields and hop-gardens which lay in its path (it caused temporary floods, and broke some of White's windows and much of his garden glass); but because only part of the parish was affected, the subsequent growth of the lacerated hops could be compared with that of hop plants which had been left untouched. Unexpectedly, and while the damaged corn never fully recovered, the hop plants which had been slashed, or one could say trimmed, by the hail produced by far the finer hops later in the year. The other conditions impinging on both the lacerated and the unlacerated hops, the soils they grew in, notably, and the way they were cultivated, were the same. White asks in conclusion, therefore, whether the top shoots of strong hop plants should not be pinched out as a matter of management by the hop growers? [20]

This was not an 'enquiry' set up by White, but it nonetheless illustrates the 'yes but' questioning and checking he regularly made use of. In *Selborne*, the process is illustrated in some of the passages in which he 'contradicts himself', or adds an abruptly qualifying piece of information in the middle of a paragraph. In one of the early letters to Barrington, he remarks that small birds are much less shy of human beings than large birds: the tiny goldcrest remains unconcerned until you are within three or four yards of it, whereas the great bustard will not allow you within three or four furlongs. He at first views this as a universal difference; he is perhaps assuming that the Author of Nature 'originally provided' large birds with a fear of man, where adding the same fear to the make-up of small birds was unnecessary. But in the same short paragraph he moves as well towards a quite different explanation – and admits quite different facts. He knows that 'in *Ascension Island*, and many other desolate places, mariners have found fowls so unacquainted with an human figure, that they will stand still to be taken; as is the case with boobies, &c.'[21] Boobies are 'large birds'. Having made the appropriate comparison – having, as I am putting it, used the boobies as a control – he now implies that the fear of man evident in a species such as the great bustard may have been learned at some time, through hard experience.[22]

Despite the sophistication of some late seventeenth-century research, the use of controls was by no means a common feature of eighteenth-century British science. Investigators of epidemic diseases had begun to use control procedures, but for the physicists of various sorts, and the chemists and anatomists, problems relating to apparatus – the technical problems involved in setting up experiments which other physicists, chemists and so on would be able to replicate – were still principal considerations.[23] White was aware of a subtly involved, multiple causation, where the laboratory scientists still thought in terms of distinct causal sequences; he saw that in any example of a behaving animal or plant, or any instance of its behaviour, numbers of possibly causal factors could overlap and interact. This might have made accurate examination impossible, but even a ready-made 'experiment' such as that provided by the hail storm showed that the difficulty could, in principle, be surmounted. Given White's patience and ingenuity, he could examine a possible cause while *masking off* the other causes and possible causes. The life-styles he wanted to explore were complex and reticulated, but – if he had the relevant records – he could not only observe but analyse even these life-styles.

2 Controls and record-keeping

White was committed to outdoor behaviour, and more than this, to the behaviour of a few creatures 'well studyed & investigated'.[24] Living things and their environmental conditions were elaborately connected; no animal or plant was an island; and his gardening apart, he seldom physically manipulated the living things he examined. (The Selborne tortoise could be hauled about, but it was a special case.) He saw that records and readings, or their manipulation, could give him special opportunities. *The Natural History of Selborne*, just as it is an example of art which conceals art, is a work of science in which relatively little account is given of the labour underlying its contents. That White was a *methodological primitive* is part of the Gilbert White mythology. He was a representative, very often, of 'lateral thinking', but was none the less 'analytical as well as observational'. His writings show us not an intellectual, but a cleverly practical, naturalist at work.

Most of his outdoor data was gathered by himself; his correspondents helped, and he made use of published evidence, but these were minor sources compared with his own observing and note-taking. The *Naturalist's Journal* is the main record of this activity. It includes a huge diversity of material; entries on wild flowers, the congregating of the swallows and martins prior to their 'withdrawal', some wine- or jam-making and a new baby born into the widely scattered White family, can all appear within a few lines of each other. The characteristic Selborne investigator can nonetheless be found within this miscellany.

Even in their original, undigested form, the successive annual accounts are not without shape or structure. While *Selborne* has appeared in more than two hundred editions, and has been translated into more than a dozen languages, the *Journal* has until recently been available to a wide public only through extracts, or in a condensed version.[25] The third of the diary/notebooks, the *Journal* runs from January 1768 to June 1793, the month of White's death. Where the *Garden Calendar* is patchy, and at best seasonal, the *Journal* is a virtually unbroken set of records. Almost without exception, the 'years' are based on the printed diary blank issued by Daines Barrington, which appeared first in 1767 and was re-published by Benjamin White in 1771. Gilbert was happy to adopt this printed schedule, although he was before long transcending its various ruled compartments. The *Journal* too is more than a factual compendium. With gardening notes continuing but now abbreviated, the annual volumes can be thought of as portraits of particular years at Selborne.[26] Some of them reflect events White could not avoid recognising, as in the 1781 volume, where the continuing drought comes to dominate all else. In others, White's own domestic doings contribute to a year's predominant tone: this happens with the *Journal* for 1777, when much of the work on the 'great parlour' was carried out, or 1780, the year of the tortoise, which was brought back to Selborne in March 1780 and was observed by the naturalist with especial pleasure throughout the succeeding summer. The entries were not all written or completed consecutively in a typical *Journal* volume. White frequently overran the printed divisions, and he inserted extra pages when he wanted to include lengthy notes, but he did not fill up pages merely for the sake of doing so. There was usually sufficient space for him to go back, where appropriate, and correct or add to what he had written at a later date.[27] Many of the touches which give this diary cum fieldbook its peculiar quality, including, most obviously, its items of national and international news, were added retrospectively and as a 'year' began to assume its particular character.[28]

As we scan a whole year of the *Journal,* a coherent picture emerges, but to examine a given year closely is to be met with large numbers of distinct entries, many of them very brief. White's succinct journal style is often rewarding: 'Sheared my mongrel dog Rover, & made use of his white hair in plaster for ceilings'.[29] But the quantitative as well as the humane aspects of the volumes reflect White's aims and preoccupations. Any recorded event is placed against a date, and the physical location of an event within the parish is often noted with as much care as the 'event itself'. In August 1790, for example,

> *Blackstonia perfoliata*, yellow centory, blossoms, on the right hand bank up the North field hill. The *Gentiana perfoliata* Linnaei. It is to be found in the marl-dell half way along the N. field lane on the left; on the dry bank of a narrow field between the N. field hill, & the Fore down; & on the banks of the Fore down.[30]

Conditions in the parish varied, and could vary. In connection with most of the thousands of entries, the temperatures and barometric pressures, and the cloud cover, wind, rainfall, snow or frost where relevant, can be checked immediately, for daily meteorological readings and notes appear almost throughout the volumes.[31] A true natural science penetrates and uncovers, we can agree; it allows us access to the unobvious. The scientific value of the *Journal* becomes clear as we recognise that relations obscured within a welter of other material may be brought to light, and just as importantly, 'inevitable' or 'obvious' relations may be falsified, by an intelligent handling of material of this kind.

In *Selborne*, the question is raised of the effects of various foods on the colours of birds and animals; the specific reference is to tame bullfinches, and the blackening which allegedly ensues if these birds are fed only on hemp seed.[32] In the *Journal,* the same idea is tried out and is firmly qualified:

George Tanner's bullfinch, a cock bird of this year, began from

> it's first moulting to look dingy; and it is now quite black on the back, rump & all; and very dusky on the breast. This bird has lived chiefly on hemp-seed. But T.Dewey's, & ---- Horley's two bullfinches, both of the same age with the former, and also of the same sex, retain their natural colours, which are glossy & vivid, tho' they both have been supported by hemp-seed. Hence the notion that hemp seed blackens bullfinches does not seem to hold good in all instances; or at least not in the first year.[33]

Where conditions have varied in the wild, and in particular where large numbers of records are available, a similar, but hypothetical, isolating and testing can often go forward. For a relevant case, we can touch again on the 'dripping' phenomenon described in the previous chapter. It seemed likely that moisture condensed on or by foliage significantly reduced the effects of a drought period at Selborne, but White could add substantiated claims to this conjecture. There was the evidence from North America, or from the regions of North America where tracts of woodland had been cut down and the levels of the streams had dropped appreciably.[34] But before this, the seeming connection between crop yields and the dripping phenomenon stood up to a testing the naturalist could conduct 'at home', using his own detailed notebooks. As with the 'control' principle itself, the case looks simple once it has been stated. Dripping had occurred regularly in the parish, and he could refer to the quantity and quality of the parish crops over a number of annotated years. These yields had fluctuated but they had not fluctuated as widely as fluctuations in the local rainfall, temperature, sunlight and so on would have led the observer to expect.

The transition from the *Garden Calendar* to the *Naturalist's Journal* took place as White's notion of himself was changing. The Garden Calendar proper ends in 1766, and the detailed letters addressed to John White, amounting almost to a course in natural history studies as his contemporaries understood this, covered the period from 1769 to 1772.[35] The correspondence with Thomas Pennant was begun in 1767, and that with Barrington in 1769; the letters in the 'Barrington' series can seem more sophisticated than those in the 'Pennant' series, but the two sets of letters were written over virtually the same period. By November 1773, stimulated as it seems by Barrington, White was writing the first of the 'hirundine papers', that concerning the house martin.

By this date, he was concentrating on 'manners' or conduct, and was viewing himself as participating in a much larger project than that of his own observing. By the same date, he was beginning to confront problems he would not have had to deal with as a mere observer. In a scientific enquiry, we said, a lengthy process of elimination may have to be gone through before a correct hypothesis is hit upon. Examinations giving White merely eliminating results took up much of his time; with several overlapping 'puzzles' – including some of the questions concerning bird migration to be looked at in the next chapter – he was still eliminating at the time of his death. Where he reported his negative findings in *Selborne*, he in several cases toned these findings down, or made them look less negative than they were; but two centuries on, these findings too can be classed with his achievements. They show us a naturalist who was working in almost virgin territory, and who, without making a great fuss about it, was using methods peculiarly appropriate to that territory.

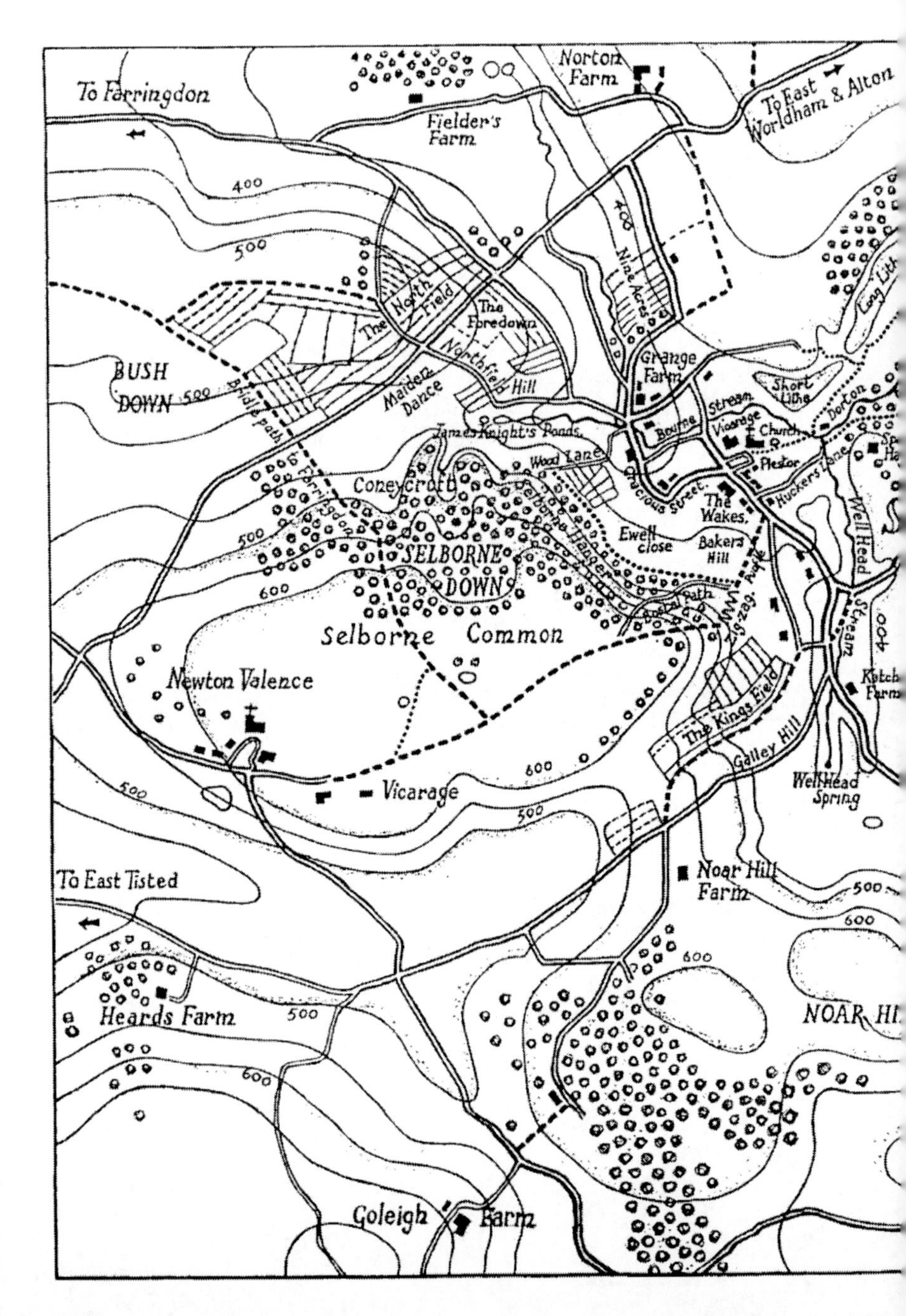

Eighteenth-century Selborne, a map built up from primary sources. The village lies 'in the shelter' of Selborne Down. The eighteenth-century parish was without metalled roads, and th ancient lanes, bridleways and footpaths reflected the contours of this 'abrupt, uneven country

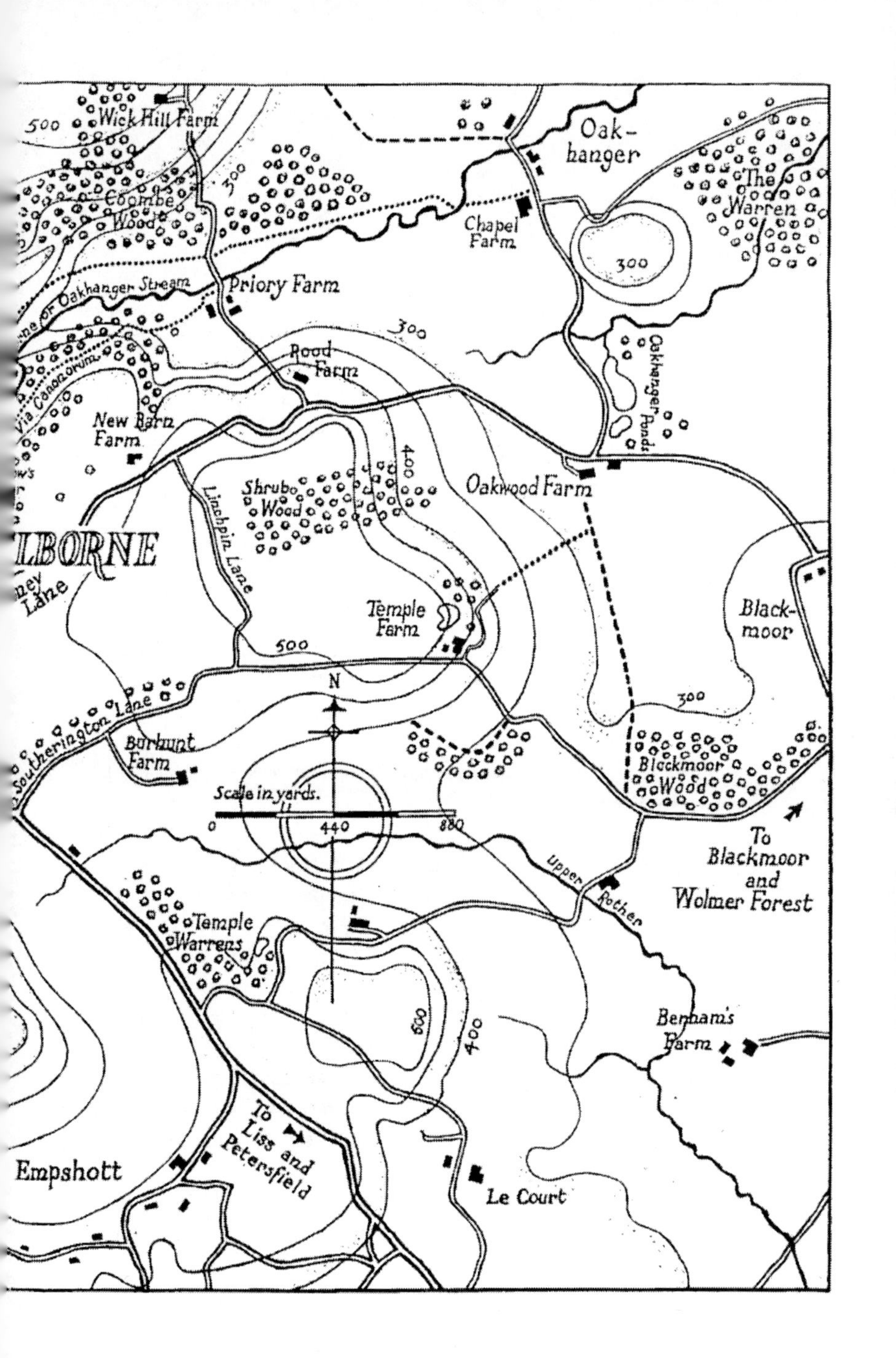

Wick Hill Farm
500
300
Oak-
hanger
The
Warren
Coombe
Wood
Chapel
Farm
300
Oakhanger Stream
Priory Farm
300
Rood
Farm
Via Canonorum
New Barn
Farm
Oakhanger Ponds
400
Shrubs
Wood
Oakwood Farm
LBORNE
Linchpin Lane
Lane
Temple
Farm
Black-
moor
500
N
300
Southerington Lane
Burhunt
Farm
Blackmoor
Wood
Scale in yards.
0
440
880
To
Blackmoor
and
Wolmer Forest
Upper
Rother
Temple
Warrens
600
400
Benham's
Farm
To
Liss and
Petersfield
Empshott
Le Court

Chapter 4

THE DISAPPEARING 'SWALLOWS'

1 Migration versus hibernation

The 'hirundine' puzzle, the question of what happened to the 'hirundines' in the course of each autumn, was no puzzle at all for many of White's contemporaries. Dr. Johnson could assure his listeners that 'Swallows certainly sleep all the winter. A number of them conglobulate together, by flying round and round, and then all in a heap throw themselves under water, and lye in the bed of a river'.[1] Johnson was hardly a naturalist, but the same view was apparently accepted by Linnaeus, and was certainly expressed by various of his students. As White remarks, one Swedish author speaks as confidently of the swallows retiring under water in September as he would of his domestic hens retiring to the hen-house at sunset.[2]

Daines Barrington, not to be outdone, held that for the duration of winter many of the 'swallows' immersed themselves in mud, at the margins of rivers and lakes. What interested White, however, was the debate as to *whether* these birds hibernated during the winter months. This debate, too, already had a long history, he appreciated. In the mid-sixteenth century, Pierre Belon had described the mass movements of apparently migratory birds, and a hundred years later Sir Thomas Browne – the physician, literary man and Norfolk naturalist – had taken the possibility of migration for granted:

> How far the Hawks, Merlins and wild Fowl which come unto us with a North-west wind in the Autumn, flie in a day, there is no clear account; but coming over Sea their flight hath been long, or very speedy. For I have known them to light so weary on the coast, that many have been taken with Dogs, and some knock'd down with Staves and Stones.[3]

In the matter of the swallows, among other 'summer birds', Francis Willughby had been unable to decide finally whether they migrated to distant regions or retired to 'lie torpid'; but he thought it highly probable that they travelled south at the decline of the year, to 'Egypt, Aethiopia, &c' and to weather more congenial to them than our winter weather.[4]

White worked on the migration/hibernation problem with application from the 1760s onwards. He found it both intriguing and frustrating; his best and even his first-hand evidence seemed to give weight first to one side of the controversy and then to the other. A fact he rated highly was that at Selborne small bands of house martins would appear during the first week in November, well after the main body had withdrawn. They would hunt and play for a day or two over the village and along the beech hanger. This favoured the hibernation hypothesis, on balance, he thought:

> Do they withdraw & slumber in some hiding-place during the interval? for we cannot suppose they had migrated to warmer climes, & so returned again for one day. Is it not more probable that they are awakened from sleep, & like the bats are come forth to collect a little food?[5]

Records of unusually early hirundine arrivals perhaps militated on the same side. A few martins or swallows might show themselves briefly in early spring, well before the arrival – or emergence – of birds of these species in their usual large numbers.

Unsurprisingly to us, White found no 'sleeping' or torpid birds, although he undertook careful searches even in his last years. He led a large party of 'young men' in combing Selborne Down for hibernating house martins, and a week later sent two labourers up to continue this work, early in April 1781. The episode is of interest in view of observations he had made during the previous October. He had been paying special attention to a large gathering of house martins which seemed to settle and roost early in the evening on the steep eastern side of the Down, among the beech and oak scrub 'above the cottages'. After watching their descent into this scrub on October 13th and 14th, he went home well pleased, he says, thinking it likely, in view of the advanced season, that their taking up this sheltered roosting place 'might lead to some useful discovery, & point out their winter retreat'.[6] (As he re-casts this incident in *Selborne*, it was one leaving him virtually convinced that the martins had established winter 'sleeping' quarters on the side of the Down.[7]) But he and his helpers were disappointed; the one martin which was seen in the village street during the second day of this search could not be offered as proof that there had been hibernating birds in the beech scrub.[8] On a later occasion he directed the thorough examination of a dilapidated thatched roof, where he had seen some young swallows gathering the previous autumn, but without success.[9] A friend of his tried to *dig out* some sand martins, which if they hibernated would surely make use of their own convenient nesting tunnels; but although this gentleman opened 'about a dozen holes', no torpid sand martins were discovered.[10]

Yet the possibility of hibernation could not be ruled out merely on account of these failures. Virgil himself had imagined the swallow 'asleep in her hollow tree'. In White's time there were innumerable extant claims concerning hibernating birds; allegedly, they had been found, sometimes in large numbers, in church towers, mine shafts, rabbit burrows and falls of rock. Early in the century, the theological naturalist William Derham (1657–1735) referred to a meeting of the Royal Society at which 'eye-witness testimony' had been provided by a Dr Colas of hibernating swallows being drawn up by Polish fishermen from deep lakes.[11] Derham himself could vacillate on the 'withdrawal' question, and in the 1740s the widely travelled Mark Catesby was referring to the hibernation accounts as 'ill attested, and absurd'.[12] However, an 'inquisitive clergyman' assured White himself that as a boy he had seen some torpid swifts taken from a demolished battlement; they revived when placed near the fire. By the 1760s, the hibernation belief was again being busily asserted – on the strength of a mass of apparently first-hand evidence.[13]

William Derham had reported the 'squeaking of swifts' in his roof well before these birds made their first appearance in his district. As White remarks, 'It is a pity that so curious a Naturalist did not proceed to the taking-down some of the tiles, that he might have satisfied his eyes as well as his hearing'.[14] But there were alternative ways if not of finally testing the hibernation hypothesis then of examining it seriously. The retirement and emergence of creatures known for certain to hibernate followed changes in local temperature with great fidelity; 'nothing so much promotes this death-like stupor as a defect of heat'.[15] Animals were not mere products of outside stimuli, and White came across instances occasionally which seemed to bring this primacy of local temperature in question, but the great preponderance of his own evidence supported it. The famous tortoise, part of whose fascination for White lay in its being a hibernating creature, retarded or advanced its coming out of the ground in spring, and its retirement in autumn, in direct accordance with weather conditions, and most particularly temperature. It (he, she?) could sometimes be observed beginning the slow task of burying itself, and then stopping as it was seduced away from the work by the return of sunshine or warm air, several times over.[16] During the autumn and winter, the local bats were especially accurate indicators of 'heat and cold', White found. Whereas the tortoise hibernated with some persistence once it was buried, although it too could reappear, the bats came out and hunted whenever the temperature rose to 50°F or above. They behave in this way, he says, 'because at such a temperament of the air Phalaenae are stirring, on which they feed'.[17] (Present-day authorities give the emerging temperature for common bats as 47°F, but they add that wind also comes into the question. If the weather is at all windy, the bats wait for a higher temperature: wind disperses and discourages flying insects, and in windy weather, therefore, only a higher temperature brings out the insects in what are worthwhile numbers for the bats.) But the rising temperature itself roused the bats from 'sleep', he assumed. The emergence or withdrawal of an animal or bird which was supposed to hibernate could be expected to show this exact tallying with local temperature, then, and local temperature could readily be measured. 'Tortoise comes forth', or, for the first swallow, 'Hirundo domestica!!!', were expressions of White's satisfaction at these events, but many of the hundreds of first and last appearances preserved in his diary/notebooks belong also with sustained enquiries. One of these concerned the possible relation of measured local temperature to the comings and goings of the hirundine species.[18]

There is nothing absurd about the idea of bird hibernation; and often, White's own records showed, the first and last appearances of certain of the hirundines – these as well as the first stirrings and then the withdrawal of bats, bees or tortoises – did show a close coincidence with temperature changes.[19] During two days of 'sudden, summer-like heat', in March 1777, creatures of several kinds were active in the Selborne neighbourhood a month before their usual time. They included 'many house swallows'; the details are given in both the *Journal* and *Selborne*.[20] This short, warm period was followed by harsh, icy weather, and the creatures all or most of them retreated, the swallows not being seen again until April 10th. When house martins appeared briefly in the parish in November, well after the main body had retired, they usually did so in warm or at least mild weather.

But the temperature connection, or supposed connection, was by no means always indicated. There were years when, even though the local weather seemed entirely favourable, the house martins, for instance, were late in making their first appearance. 1774 was such a year, and White remarks in the *Journal* that 'these long delays are more in favour of migration than a torpid state'.[21] When the house martins did return in numbers in 1774, early in May, they were exceptionally numerous, we also discover from the *Journal* notes; their non-appearance the previous month had not been due to a fatality of some kind. On other occasions, the house martins did not 'reappear' over the village in November, even though the weather remained 'summer-like' and the naturalist was looking out for these birds.[22] Even more pertinently perhaps, in their first arriving or emerging in spring, the swallows and martins would every now and again *defy* what ought to have been discouragingly – or inhibitingly – *cold* weather. In April 1784, in a letter to a favourite niece (who had recently married), Gilbert asks, 'Pray how is it, Mrs. Mary, that the present most *bitter* spring does not at all retard the coming of the summer birds?'[23]

Overall, White's records showed that hibernation could virtually be ruled out, as an explanation of the comings and goings of the hirundines. (An exact correlation with local temperatures would not have proved that these birds were hibernators, but the widely varying, *non*-correlating pattern of records here did effectively prove that they were *not* hibernators.) The evidence is not set out fully in *The Natural History of Selborne*.[24] For Gilbert White too, it becomes clear, the hirundines were a special case. In one instance, material relating to the disappearance of these species in autumn, though included in *Selborne*, is rearranged, so that its true significance is diminished. White refers to the unsuccessful search of the beech scrub on Selborne Down, conducted in the spring of 1781, in his book, but he refers to it well *before* he enlarges on the events recorded in October 1780, the roosting of large numbers of house martins in the beech scrub – which caused him to make the search. The accounts are rearranged, in the published work, and no mention is made of the links between them.[25]

The hirundines, as White discusses them, can become a special case for us. They can obscure what he has to say about a whole range of other migratory species, work often contrasting with his treatment of the hirundines. Unlike some of his contemporaries, he did not speak as if the 'swallows' were the only birds which could be seen in Britain for only part of the year, and did not generalise from these to other periodic species. In his first letter to Barrington, indeed, he lists *thirty seven* species as migratory, or as he terms them 'passage', birds at Selborne – and includes all the hirundine species among them. If there are some unexplained appearances and disappearances, 'migration in general' is undoubtedly a fact, he insists in a subsequent letter. He had been receiving eye-witness information from John White at the Gibraltar station, and he summarises some of this for the reluctant Barrington. John – like 'old Belon' – personally observed huge movements of birds 'for many weeks together, both spring and fall', and during these weeks, 'myriads of the swallow kind traverse the Straits from north to south, and from south to north, according to the season'. These vast processions, he goes on, 'consist not only of *hirundines* but of *bee-birds*, *hoopoes*, *oro pendolos*, or golden thrushes, &c. &c. and also many of our *soft-billed summer birds of passage*; and moreover of birds which never leave us, such as all the various sorts of hawks and kites.'[26] Here we are suddenly quite free of the anecdotes and traditional explanations.

Considering 'summer birds', as opposed to merely the hirundines, White looks critically at the view that the smaller withdrawing species, anyway, are *too weak* to make journeys across continents and oceans. (Because of this 'weakness', Barrington had doubted whether migration took place regularly among any of the more familiar summer species.) That the hirundines might be physically incapable of such travelling can be summarily dismissed, he sees. Comparatively speaking, swallows and swifts, if not the two sorts of martins, are particularly well equipped to make long journeys.[27] In some instances, the 'perils' of migratory travelling can themselves be exaggerated; if we think about it, 'a bird may travel from *England* to the equator without launching out and exposing itself to boundless seas, and that by water at *Dover*, and again at *Gibraltar*.'[28]

Many of the other summer birds are short-winged, he says, and seem to flit rather than fly, but regarding these species we have *no* reports of their being found in nooks and crannies in a torpid state. In spring, moreover, these too can sometimes make their first appearances with us in thoroughly inclement weather. In April 1790, redstarts, blackcaps and spotted fly-catchers returned to the Selborne parish although five inches of snow lay on the ground. These were undoubtedly migrants, and White asks in wonder: 'how could they, feeble as they seem, bear up, against such storms of snow & rain; & make their way thro' such meteorous turbulencies as one should suppose would embarrass & retard the most hardy & resolute of the winged nation'?[29] Comparing the swallow kind with these flitting species, he remarks to Barrington that, 'entre nous', the departure and return of the short-winged summer birds provides the real avian mystery.[30]

Even among the hirundines, there had always been one bird which, at least in its disappearance, quite contradicted the supposed connection with local temperature. The swift almost invariably withdrew, or departed, for White could never seriously think of the swift as a hibernator; it might leave Great Britain in the middle of August during 'our sweetest weather', this is to say, and it apparently removed at an early date even from Andalusia.[31] The swift among other species raised the question of the 'purposes served' by bird migration, therefore. White assumed – or assumed whenever he addressed the question as a naturalist – that whatever might immediately trigger migratory behaviour, an important cause of these impressive annual movements was fluctuation in food supply; the main purpose served was that of allowing the birds concerned to find sufficient food. But how could the – late summer – departure of the swift relate to a food shortage? And if it did, how were the other hirundine species able to remain in 'this country' for two months longer than the swift? They stayed until the middle of October, and normally reared two broods, where the swift reared only one.

White's tentative answer drew on the fact that, if the swift was a relative of the swallows and martins, in several respects it was also an exception among the hirundine species.[32] Both its mandibles and its feet were peculiarly formed, and typically, it worked *higher in the air* than the 'true swallows'. That this high flying and hunting might have a bearing on its early departure occurred to White, he says, as he noted the early 'retirement or migration' of his newly discovered 'great bat' (the noctule bat), and as he recalled its working at a higher altitude than the common bats. Perhaps the swift, at its own height, took its own sorts of 'gnats, scarabs or *phalaenae*', and perhaps the insects taken by the swift were active for a shorter period than those taken by the other hirundines.[33]

The function of migration, the question of what migration did for a migratory bird, continued to occupy White. Objections to guaranteed food supplies as the answer, or whole answer, could be raised in the cases of several birds in addition to the swift, and with some of these the objections could not be disposed of easily. His commonest winter migrants, the red-wings and fieldfares, for example, depended on much the same food as their 'congeners' the blackbirds and thrushes, during their period in Hampshire, and large numbers of blackbirds and thrushes remained in Britain all the year round. If food was the decisive factor, why did the red-wings and fieldfares leave us? Could they not subsist here happily on the same food as the blackbirds and thrushes during the summer months? If the winter migrants do not throw doubt on 'food' as an explanation of migratory behaviour, he continues, they throw doubt on 'food alone' as an explanation.[34]

He is approaching an answer to this further difficulty as he recognises that 'birds may be so circumstanced as to be obliged to migrate in one country and not in another';[35] migration is not as fixed a characteristic of the 'migratory species' as it might seem; and as, he assumes, regarding fieldfares and migratory woodcocks, anyway, that the districts to which they proceed once the winter is over must, somehow, suit them better as breeding localities than do the British Isles.[36]

He understood that thinking about the birds of his parish *merely as Selborne birds* could leave him seriously disabled, if not as an observer then as an investigator. Members of some of the species we think of as winter or summer migrants may be found in this country, in small numbers, in *summer* and *winter* respectively; he mentions the wheatear as one example. The choice of this species may surprise the present day bird-watcher, but White's essential point is well made; some birds though they make a southern migration in autumn, get no further than, for instance, our Channel coast.[37] In addition, and depending on the season, he sees that many of the birds which are mere visitors to a region such as the Selborne parish are of the same species as some of the residents, and are therefore easily confused with those residents. He suspects that large-scale 'internal migrations' take place within Great Britain each year, and that these directly affect bird numbers in his district. The huge flocks of chaffinches present at Selborne during, but only during, the winter, and the prevalence of female birds within these flocks, are perhaps the result of an internal migration; only, 'I want to know, from some curious person in the north, whether there are any large flocks of these finches with them in winter, and of which sex they mostly consist.'[38]

White could sometimes compare his own 'first appearance' records with those of observers in other parts of England. In 1775, when the house martins were again late in their spring arrival at Selborne (despite the more or less regular behaviour of the other 'summer birds'), he could add that the same thing had happened at Blackburn, where his brother John eventually settled, and at Lyndon in Rutland.[39] In 1774 he could compare his own dates with those for both Blackburn and South Zele in Devonshire (though whether he was provided with the relevant weather reports, as well as the dates, he does not say).[40] He realised that the fundamental migratory movements were trans- and inter-continental, and that most of the migrant species he could observe were representatives of these movements. This total ebb and flow of birds would have to be allowed for, in any attempt to understand migration adequately. No one investigator could experience entire migratory patterns; on the contrary, methodical, co-ordinated work by observers placed along what seem to be the migration routes would be necessary, if real progress was to be made. The 'gentlemen of the navy' might be co-opted. They could keep notes of what birds alight on their ships, 'at what seasons, in what latitudes, and in what weather, and from what points', and thus 'trace them in their very course'.[41] Birds take the shortest way across water, White thinks, and records kept at what seem to be the bottlenecks – or rather a 'comparative view' of these records – will be important.[42] He did not envisage the internationally organised ringing and radar tracking programs which are essential aspects of the modern study of migration, but had he been able to envisage them, we can say, he would easily have recognised their value.[43]

When present-day observational records are gathered together, and when the annual movements of swallows, for example, over the total migratory range of this species, from central Africa to the Arctic Circle, are plotted along with climatic events, their progress within this range does show a relationship with changes in 'heat': according to Eric Tomlin, they 'keep almost perfect step' with a temperature of 48° F.[44] They advance northwards in spring as a rising temperature allows them the insect food they need and as, therefore, their various nesting localities again become usable. Food, sufficient not only for the returning birds but also for *the broods of young they will rear once they are re-established,* is the key to the spring migration, naturalists now accept. Over great periods of time, climatic changes have been an influence on bird migration. Without winters cold enough to freeze soil and water, and to greatly or entirely reduce insect activity, much of the southwards migration observable in the northern hemisphere each autumn might never have begun. But the great return journey, the return in the course of (our) spring to the usual breeding places, is a reflection also of bird 'dispersion': it is part of an annual anticipation of the fact that soon there will be many more mouths to feed.

Other local naturalists kept more or less careful records, but White adumbrates the approach to migration within which record-keeping can be constructive. He did not get to the crucial matter of mating and then feeding a family, although dispersion interested him, as we shall see, and although he observes that 'love' and 'food' are the great regulators of animal conduct.[45] The grasp he did achieve – and his care in delimiting what he could legitimately claim – appears from a letter to Thomas Pennant. Referring again to the evidence sent by John White, virtually his only dependable evidence here, he says that if species we know as summer birds

> are found in *Andalusia* to migrate to and from *Barbary*, it may easily be supposed that those that come to us may migrate back to the continent, and spend their winters in some of the warmer parts of *Europe*. This is certain, that many soft-billed birds that come to *Gibraltar* appear there only in spring and autumn, seeming to advance in pairs towards the northward, for the sake of breeding during the summer months; and retiring in parties and broods towards the south at the decline of the year . . .[46]

Gibraltar he describes as a 'great rendezvous, and place of observation'.

He knew that during any particular day, or for that matter month, *prevailing winds* could advance or retard migratory progress (where appropriate, wind direction, often with a rough indication of strength, is one of his daily weather records): woodcocks, for instance, may be driven quite out of their usual routes by a storm.[47] He could show authoritatively that the northward advance of the swallow kind might temporarily not keep step with 48° F, or any other temperature which can be agreed on beforehand. But, checking his records, we find that the maximum temperatures which go with his first appearance dates are very frequently in the range of 46°–50° F.[48]

2 The attractions of hibernation

The 'imbecility', or weakness, of small birds is often illusory, White says. When necessary, they can travel much further than might be expected, and on the other hand, the tiny goldcrest, a resident in Britain, survives our coldest winters so long as it can find food. (If vulnerability to cold is one reason, it 'seems not to be the only reason' why some small birds leave our shores in autumn.[49]) And yet it is to imbecility that he returns in the case of the hirundines. When he narrows his focus and concentrates on them alone, he remains determined to accommodate hibernation as a possibility, or rather, hibernation and migration as both possibilities. He implies sometimes that some species of swallows hibernate and some species migrate, and suggests alternatively that while the *adult* birds (perhaps of all the hirundine species) travel south for the winter, the *young* birds, and more especially the young of the second broods, which may have been out of the nest for only ten or twelve days at the time of the departure of their older relatives, remain and hibernate in the country of their birth for at least one season.[50] These newly fledged swallows and martins cannot be supposed to make the migratory journey – on account of their physical frailty.

White could instigate a search for torpid birds even in his seventy-third year.[51] Regarding the hirundines and their withdrawal, he received misleading and simply false information, both from 'common report' and from sources which ought to have been trustworthy; and, until the end of his life, he could himself evade facts on the hirundine questions. We can recall his approach to the ring ouzels here, his 'new migrants', as he called them. He assumed the November house martins, which appeared in the parish well after the withdrawal of the main body of these birds, to have re-emerged, although he reacted to the periodically but briefly sighted ring ouzels quite differently – and much more shrewdly. It is now obvious that the November house martins appeared at Selborne for a few days while *passing through* the locality. They had come from somewhere far to the north of the Selborne parish (they had not spent the summer in the parish). This was a possibility White largely ignored – although in the case of the ring ouzels, his first hypothesis was that the birds were already *en route*. As he ascertained in spite of Pennant, the ring ouzels used the Selborne downs as a temporary resting and 'bating' place.[52]

A wider, cultural question underlay the migration–hibernation debate. Hibernation seemed consistent with a stable and divinely planned natural world, where migration – the business of braving the elements annually and removing from one region to a perhaps far distant region – seemed to imply a hand to mouth expediency, a lack of adequate planning, in this department of nature. White could challenge the 'planned and stable' account, as we shall see, and he admired the adaptability and enterprise, as he sometimes thought of it, of migratory birds; but he seemingly remained attracted by the image, the mental picture, of hibernating creatures. Here, I suggest, we again have evidence of his 'multi-levelled' personality. He notes at one time and another that hedgehogs, carp, harvest mice, 'water rats', slugs, snails and insects of many kinds doze away the cold and uncomfortable months. Bats and tortoises do so, and the idiot bee-boy described in the *Natural History* became more or less torpid in winter. Even cats and dogs, White says, have 'a power of accumulating rest', and sleep much of the time when the weather discourages outdoor activity.[53] The naturalist-curate himself remained active and conscientious throughout the 'cold and wet season'; but during this season, and at one level, he took pleasure in the thought of these snug and well organised 'sleeping' creatures.

In the case of the swallow, a beauty he was always aware of was added to the special aura surrounding this bird.[54] His feelings for the 'swallow kind' – he perhaps hoped subconsciously that they would not divulge their essential secrets – appear in a note such as the following, made towards the end of his life:

> The congregating flocks of hirundines on the church & tower are very beautiful! When they fly-off together from the Roof, on any alarm, they quite swarm in the air. But they soon settle in heaps, & preening their feathers, & lifting up their wings to admit the sun, seem highly to enjoy the warm situation. Thus they spend the heat of the day, preparing for their emigration, & as it were consulting when & where they are to go.[55]

Although here, we notice, he is again assuming that these birds will set out on a journey.

Chapter 5

BEHAVIOUR: BIRDS AND OTHER ANIMALS

1 The 'life and soul' of natural history

In discussing the disappearing hirundines, White could allow 'analogous reasoning' an excessive influence, and could arbitrarily separate the Selborne hirundines not only from other summer birds but from hirundines in other places. Both were the sorts of mistakes he also exposed, and as a rule he quite avoided both in his own work.

He did not isolate species merely in the process of writing about them. *The Natural History of Selborne* is quite different in conception from books concerned mainly with identification, in which species are summed up one to each page or brief section. Until well into the twentieth century, such books depended heavily on 'descriptions', or more or less detailed accounts of physical appearances. Descriptions were assumed to facilitate identification – and were all that was thought necessary for the ascertaining of 'specific differences' – but they could be taken quite conveniently from preserved and mounted specimens.

'Describing' and collecting were related, and White was fully aware of collecting, of course. The alliance of collectors who were willing to have rare animals and plants hunted virtually to extinction with professional classifiers and phylogenists was still to come, but White tells of being shown some 'astonishing' collections of 'stuffed and living birds' in London by Daines Barrington. He noticed immediately the preponderance among the living birds of seed-eating species – or species which could be kept or fed easily.[1] Collections of live animals, he implies, may be as misleading in their way as collections of dead ones.

Selborne is largely distinct also from the county 'natural histories' produced in Britain during the eighteenth century. These all owed something to Robert Plot's *Natural History of Oxfordshire* (1677) and *Natural History of Staffordshire* (1689). They were local or regional surveys, but in the main were regressive in conception. In *The Natural History of Cornwall* (1758), William Borlase lumps starfish, polyps and medusas in with his 'water insects'. Like the other county naturalists, Borlase considered wild animals and plants chiefly in terms of their colours and structures, and where a species had been described already, in some other work, he could leave out even this information. Robert Plot had seen no reason to elaborate in his *Staffordshire* on those birds and 'quadrupeds' which could be found in that county but which he had covered already in his *Oxfordshire*. Its surroundings made no significant difference to an animal or plant, he seems to have assumed. In the same spirit, Borlase refers to only five or six of the birds to be found in Cornwall, and provides details of only one, the 'Cornish chough'. The others, he says, have all been adequately dealt with in Willughby's *Ornithology*.[2]

Local conditions interested the county naturalists, but principally as these affected the human inhabitants. As much as anything else, the county histories were instructive handbooks for landowners and other potential investors.[3] William Borlase provides details of the mining and smelting industries of eighteenth-century Cornwall, and backs these up with a serious attempt to understand the geological, or mineralogical, character of his region. The same writer offers almost nothing concerning animal lives, and seems to assume that if he had involved himself with, for instance, the lives of quadrupeds, this would have given him very few problems. He does not regard animals as mere machines, and he condemns any readiness 'to give them pain wantonly'.[4] He refers to the marginal variations in shape and colour which can occur within some quadruped species; they differentiate individual animals 'as much as the human features distinguish mankind one from another'. But he goes on:

> Wherefore then was this variety bestowed upon brutes? Are they at all sensible of such diversity? Are they more happy, or more useful to one another for it? No. This variety then is doubtless intended for the sake of man, to prevent confusion and decide and ascertain his property.[5]

White refers to Robert Plot on several occasions; his notion of study in one locality may have been suggested originally by the writings of Plot; but in the event, White and the 'county naturalists' have little other than their book titles in common.[6] Every country and province 'should have it's own monographer', White says,[7] but for him the habits of animals and plants observed in the field were 'the life and soul' of natural history. Explicitly, 'conversation', or personal intercourse of all sorts, was integral to this subject matter. White seldom disagreed with those who assumed the colours and shapes of wild animals and plants to be fixed, but he could express and justify serious doubts about the fixity of behavioural adaptations. The Selborne parish, though it extended to eight or nine square miles in his time, was small compared with a county, but it was particularly varied; and the limited help he got from his correspondents – who at their most useful represented geographical regions quite different from his own – was put to intelligent use.

2 Personal relations and communication

White's phrase 'life and conversation' is used only in zoological contexts, although, as we shall see, the same concerns dominated his botanical work. (He could examine plants very closely, but he made little or no contribution to systematic phytology. He is a counter-weight to Linnaeus, or Linnaeus in the work for which he is most celebrated.) Animal sociability greatly interested him and, along with this, a periodic 'jealousy' which seemed to contradict this sociability. In certain circumstances, the jealousy could replace 'even parental affection'.

References to a local rookery and notes on the behaviour of the rooks, appear from time to time in White's journals. He passed the rookery on the weekly trips he made to Faringdon for many years, and clearly, on these occasions, he gave himself time to observe the goings-on of the rooks. Though essentially gregarious, these birds were quarrelsome during the time of nest-building, he found. 'Some unhappy pairs are not permitted to finish any nest 'til the rest have completed y^{ir} building; as soon as they get a few sticks together a party comes & demolishes the whole'.[8] Even if one of the pairs thus discriminated against tries to build apart from the main colony on a 'single tree', the nest is at once destroyed; although, as White notes elsewhere, a pair of house sparrows may be allowed to build in the fork under a rook's nest.

He offers no explanation for this 'quarrelling', remarking only that it seems 'inconsistent with living in such close community';[9] but various forms of behaviour having much in common with it are pondered to good effect. The affectionate behaviour of, for instance, a domestic hen towards her young ones is succeeded by a jealous, almost spiteful, attitude. She may guard the small chickens with the greatest solicitude, and may supervise them as fledglings until they can fend for themselves, but once they reach this self-sufficient stage, she will turn on the young ones and 'cruelly drive them from her'.

This reversal is typical of the 'feathered race', White thinks; and in one form may have an immediate use and value. In September 1778, 'The young house martins of the first flight are often very troublesome by attempting to get into the nest among the second callow broods; while their dams are as earnest to keep them out'.[10] But it also has a more general function. Were young birds allowed to establish themselves wherever they liked, 'one favoured district would be crowded with inhabitants, while others would be destitute and forsaken'.[11] That a stabilising mechanism of some sort is at work is clear from the fact that Selborne itself does not become over-crowded. If forty pairs of house martins – and many more than this breed in the parish – each bring up two broods of young, and if there are four young birds to a brood, by the end of the season, and including the parents, the result will be four hundred birds, but the numbers of house martins departing from the parish during a particular autumn bear no relation to the numbers breeding there the following year. In the case of the swifts he can give exact figures, because their nesting sites are few. Swifts are not prolific breeders, producing only one brood during their short stay and 'only two' young ones to a brood; barring accidents, therefore, they double their numbers in the course of each summer. Yet eight pairs, and never more than eight pairs, nest in the village annually.[12]

Many of the increase must die or be killed before the next nesting season, but White is interested here in what birds as it were in possession of a congenial district do in the face of newcomers, or the effect of what they do. In the case of the jealousy which replaces parental affection, it helps 'spread birds across the countryside', he says. This parental behaviour links up, therefore, with the 'rivalry and emulation' which male birds can exhibit towards each other during the breeding season. White lays it down as a maxim in ornithology that 'as long as there is any incubation going on there is music',[13] and the rivalry, as he draws attention to it, is not over females but over what was later to be called 'territory'. During the breeding season the male birds 'can hardly bear to be together in the same hedge or field', and by their singing and aggressive mien, proclaim their authority over specific, limited 'cantons'. Maintaining their territories – and though he does not reach a comparable inference concerning migration – they are able to pair and rear young, and perhaps more than one brood of young, with every chance of success. White was not the first to record territorial behaviour, but he was one of the first to appreciate its biological significance. Were this 'dispersion' not to take place, he implies, the species themselves would be endangered.[14]

A cock bird asserts his claim to the area he has defined by his singing and posturing, but this conduct is efficacious only in so far as birds which might have been in competition with the cock and his mate acknowledge this show of defiance – or react to it by keeping off. Communication between animals was a crucial aspect of their 'conversation', White saw.

Selborne includes a brief but thoughtful paper on bird songs and calls. Even the calls and refrains we think of as delightful or amusing are by no means ends in themselves, White insists. If birds can be said to use languages, these are ancient and elliptical languages; 'little is said, but much is meant and understood'. The calls and songs of birds may be expressions of their various 'passions and wants'; he refers to 'anger, fear, love, hatred, hunger, and the like'.[15] The notes and cries of a farmyard hen with chickens are consistently different from those of a childless hen, and hens with or without young ones are transformed again, and will 'upbraid, execrate, insult and triumph', as they attack a dangerous enemy which has been placed at their mercy. He thinks of the calls and songs as signals. If they indicate arousal, in many cases they also elicit responses from other birds, of the same or even quite different species. A pullet which has just laid an egg emits a joyful clamour, which the cock and the rest of his mistresses immediately repeat and which is taken up by any other cocks and hens within earshot. The members of the whole extended flock, it seems, proclaim their solidarity with the layer. A male swallow, 'by a shrill alarming note', gains the attention not only of other hirundines but of other small birds, and alerts them to the presence of a hawk; a co-operative buffeting and confusing of the hawk will ensue, with the result that he may be quite driven from the village. Gregarious birds which shift their quarters during the hours of darkness keep up a constant calling, the naturalist notes again, which prevents them losing their companions. The repeated short, quick notes of the stone curlews which pass over his house each night during the summer months are 'signals or watch-words to keep them together'.[16]

White establishes that the three leaf warblers are distinct species with distinct songs, but he does not dwell on the importance of their respective songs for the birds in question. (The songs seem to enable these otherwise similar birds to recognise and mate with members of their own species.) He appreciates, though, that communication does not depend only on sounds or hearing. Many animals recognise and are recognised by means of smell, and to this extent, at least, can be said to send and receive olfactory messages.

> After ewes & lambs are shorn there is a great confusion & bleating, neither the dams nor the young being able to distinguish one another as before. This embarrassment seems not so much to arise from the loss of the fleece, which may occasion an alteration in their appearance, as from the defect of that *notus odor*, discriminating each individual personally: which also is confounded by the strong scent of the pitch & tar wherewith they are newly marked; for the brute creation recognise each other more from the *smell* than the *sight*; & in matters of *Identity* & *Diversity* appeal much more to their *noses* than to their *eyes*.[17]

White learns from sheep which have been temporarily *de*odorised, or altered as to their odours, the importance of odour and the sense of smell for mutual recognition among sheep. In a later note, he remarks that dogs too provide evidence of this olfactory communicating: perhaps he had again been 'shearing' his mongrel dog.

But visual – colour and shape – indicators are important in the lives of many birds, he agrees elsewhere. Birds are 'much influenced in their choice of food' by colour;[18] and in many bird species, colours are 'the chief external sexual distinction' they depend on. Most juvenile birds are dowdy, or quite without the colours and markings of adult birds, 'because they are not to pair and discharge their parental functions till the ensuing spring'. This retardation saves them from the advances of mature birds before they are ready for sexual involvement, or perhaps from sexual involvement at the wrong time of the year.[19]

Communication by means of 'languages' overlaps, then, with communication by means of what might be called *involuntary semaphore*. In various quadrupeds, horns and shaggy manes, beards and brawny necks, are an effect but also a proclaiming of masculine maturity. It may even be true that 'insignia' such as these influence not only females and other males but the mature male himself – the male exhibiting the insignia. The conduct of a 'fierce and venereous' boar was quite altered when its tusks were broken off; it went into a sexual decline; although, the 'candid' reporter adds, he has taken this story from a book on animal husbandry.[20]

Courtship songs and calls are touched on only superficially, but courtship display, often contrasting with a creature's usual behaviour, is described in several of White's accounts. Some male peacocks he could watch put on their glamorous performance 'when the hens appeared', and woodpigeons, though they are strong and rapid flyers at other times, 'hang about on the wing in a toying and playful manner' during courtship and as a courtship signal. (Anyone might have noticed this 'toying on the wing' in woodpigeons, but White asked what it was for.) The greenfinch exhibits such 'languishing and faltering gestures' in the course of the same relationship that it can be mistaken for a dying bird, and the cock snipe forgets his normal flight style and 'fans the air like a wind-hover' while courting. The snipe's peculiar 'bleating' or 'humming' is heard only during the breeding season, White thinks. He inclines to the view that this 'bleating' is vocal, but he adds from his own observation that it is produced only during the bird's rapid descent.[21]

Danger signs, provided by a predatory species inadvertently and made use of by other species as they defend themselves or try to escape, are referred to in passing, for instance in the paper on aural languages. The tiny shape of a hawk high in the air is recognised by a mother turkey; she watches the hawk with steady concentration and announces the danger with a special 'inward moaning'. White does not tell us whether the turkey chicks react to the shape of the flying hawk, but he writes concerning a four or five day old chicken; 'hold it up to a window where there are flies, and it will immediately seize it's prey, with little twitterings of complacency; but if you tender it a wasp or a bee, at once it's note becomes harsh, and expressive of disapprobation and a sense of danger.'[22] We cannot doubt that White had himself tested a chick in this manner.

He seems to assume that the chick has inherited this ability to respond immediately and (for it) appropriately to the black and quietly buzzing insect on the one hand, and the larger, yellow and black and more noisily buzzing insect on the other; but what he is saying first is that the chick's reaction is at once expressive and adapted. Properly, he implies, we begin to understand an item of behaviour only as the conditions, animate and inanimate, in which the behaviour normally occurs, are known and only, therefore, as the 'purpose it serves' is to some extent appreciated.

3 Self-contradiction and identification

White points to numerous instances in which behaviour seems to contradict structure, in the creatures he is familiar with. We should be outdoor observers of animals and plants, rather than collectors or mere anatomists, because structure does not give us sufficient information regarding habits, and because, more importantly, a student depending only on structure – or physique – must sometimes be entirely misled regarding habits. White could write admiringly of structural features: in his second and longer note on the peregrine falcon, in *Selborne*, he lists the physical respects – powerful chest and thigh muscles, a tearing beak, 'armed feet' and so on – in which it is 'wonderfully formed for rapine'.[23] He knew that some physiological changes directly affected behaviour. Castration, most notably, could bring 'man, beast or bird' to a close resemblance of the opposite sex; he instances domestic capons, which grow only pale combs, walk 'without any parade' and gather small chickens under them in the manner of hens.[24] But he saw too that behavioural adaptation might as it were transcend structural adaptation. Most remarkably perhaps, among both plants and animals, 'congenerous' species in many cases lead quite different lives. Comparison is as integral to his reports as observation itself. Species which are physiologically close may diverge widely as to their propensities and usual conduct, and in the environmental conditions they accept.

The hirundines provide examples of this divergence, even with the swift left out of account. They are 'closely related' species, or are very similar physiologically, but their ways of nesting vary greatly. The swallow and house martin make roughly comparable nests on or in buildings, but the sand martin, the smallest of the hirundine species, nests in tunnels excavated in sand cliffs and generally keeps quite away from human habitations. Thinking about the labour the sand martin's tunnelling requires, White observes:

> Perseverance will accomplish any thing: though at first one would be disinclined to believe that this weak bird, with her soft and tender bill and claws, should ever be able to bore the stubborn sand-bank without entirely disabling herself: yet with these feeble instruments have I seen a pair of them make great dispatch: and could remark how much they had scooped that day by the fresh sand which ran down the bank, and was of a different colour from that which lay loose and bleached in the sun.[25]

Blackbirds and thrushes are the 'congeners' of red-wings and fieldfares, but the first two are residents in Britain and the second are winter migrants only. (As far as White could learn, fieldfares did not breed even in the wilder parts of Scotland.) Red deer and fallow deer are related species, but in White's experience they keep to quite different environments and foods. Until the time of his young manhood, red deer had lived on Woolmer Forest in considerable numbers, and though they could have found little other than whortle and ling to eat on this 'sandy waste', they never crossed into the grassy glades of the adjoining Holt Forest. Similarly, herds of fallow deer inhabited the Holt, and were never seen on Woolmer. The two species of deer kept up these ecological preferences, he adds, even though the two forest tracts are separated 'only by a common hedge'.[26]

The slug and the snail are congenerous, but the 'naked slug' seems better able to cope with the cold than the 'housed snail'. Remarkably, the slug is much less willing to retire and hibernate than his – seemingly – better protected cousin.[27] The three cricket species White studied belong in this series, whatever their other attractions for him. (Insects are considered in a later chapter, but we too should avoid a merely stolid compartmentalising.) The three crickets are congenerous, but,

> while the *field-cricket* delights in sunny dry banks, and the *house-cricket* rejoices amidst the glowing heat of the kitchen hearth or oven, the *Gryllus gryllo talpa* (the *mole-cricket*), haunts moist meadows, and frequents the sides of ponds and banks of streams, performing all it's functions in a swampy, wet soil.[28]

The house cricket has the best of it, clearly, for in its circumstances winter is another summer and food and drink are always available. House crickets are 'always alert and merry', therefore, unlike the other two species. As White notes for the first time in November 1789, 'these domestic insects, cherished by the influence of a constant large fire, regard not the season of the year, but produce their young at a time when their congeners are either dead, or layed up for the winter'.[29] House crickets, this is to say, breed in all or any seasons.

The outdoor observer, or with the house crickets the kitchen observer, able as he is to ascertain habits and predispositions, can make qualified predictions regarding behaviour, where anatomy may prove quite inadequate as an aid to prediction. The 'physical weakness' of the young house martins ought perhaps to have ensured that they made no lengthy migratory journeys – but it did not. In the pursuit of natural history, logic must be subservient to observable fact; and considering the young martins, and untypically, Gilbert White was himself led astray.

Physique is not always the indicator we might expect concerning habits and habitats; and the idea of consistency among living things may be a fallible guide; we cannot confidently infer the responses of one species merely from our knowledge of some other species. The paradigm warning is given us by the cuckoo, White says, or by its willingness to leave its eggs to be incubated and its young to be reared by other birds. So exceptional is the cuckoo's conduct that if it were recounted of 'a bird of Brazil' we should refuse to believe the story; and yet, discussing the matter with Barrington, he finds that a further extraordinary fact has to be added to that of the cuckoo's parasitism. Female cuckoos seem to utilise as hosts for their eggs and young ones only birds or species able, in due course, to provide the young cuckoos with suitable food.[30]

Considering a creature such as the cuckoo, one could be forgiven for doubting whether nature is ultimately rule-observing, White continues. (A 'universally operative' maternal instinct seems to be quite subverted here.) This is not what he believes, although the field work necessary to the uncovering of rules may well be unending. He takes up the idea that the cuckoo might not deposit its eggs indiscriminately in the letter in which it is proposed. He lists the species which to his knowledge have received the attentions of the cuckoo – and finds they are all predominantly insectivorous. The wagtail, hedge sparrow, whitethroat and meadow pipit are named by him as typical hosts.[31]

He saw that the workings of migration would begin to be exposed only as a result of organised collaboration; his zoological studies evolved, but his belief in team work was explicit from the 1760s onwards.[32] While preparing a history of the stone curlew, he enlisted as informant 'Mr Woods of Chilgrove', a friend and connection by marriage, who lived on the Sussex downlands and was a 'proper person', therefore, to participate in a study of this bird. In one of the resulting papers, White makes one of the first references in print to protective colouration. The tints and markings of a stone curlew chick are such that, crouching motionless on open, flinty ground, it is virtually invisible to predators – or observational naturalists.[33]

He gives no other examples of protective colour patterns, following from this; in the *Journal*, he dwells on the peculiarly streaked plumage of a bittern, but without noticing its especial value as camouflage. Habits, however, are frequently found to facilitate survival in particular conditions, and he notes that the 'skulking' or crouching response found in many ground-nesting birds, adults or young ones, may conduce to their relative safety, even though, as sometimes happens, the crouching bird can be clearly seen. Stooping hawks (or falcons) cannot attack a bird which refuses to leave the ground, he says, and which will not even 'run before them'.[34]

His paper on the nightjar was unfinished at the time of his death, but we have some of his materials for this paper. To begin with, the nightjar would have been cleared of the 'superstitious crimes' still attributed to it when White was studying the species. There was nothing to justify the claim that the bird (*Caprimulgus*) sucked goats, he was sure, and the disease occurring along the backs of young cattle was caused not by the ill-fated nightjar but by grubs of the warble fly. (A genuine countryman, White had seen these grubs squeezed out by hand.) He could quote a farm worker who had stripped dead calves suffering from the disease, and a friend in Cheshire had studied the warble affliction.[35] He confirmed that the nightjar – or 'fern owl' or 'churn owl' – was a summer migrant, and he was satisfied that its churring note was produced vocally. In this he was again correcting Pennant, but he had watched a nightjar calling while at rest.[36] The bird could be observed during only two hours out of the twenty four, the hour after sunset and the hour before sunrise, but he could offer eye-witness descriptions of its 'glancing' agility on the wing and its way of catching 'solstitial chafers'. While in flight, it takes these insects with its *feet*, or rather with *one foot*, he says, and then, using claws and beak, picks them to pieces as it flies along.[37] His suggestion as to the purpose of the nightjar's pectinated claw is no doubt mistaken; this is now recognised as a preening claw; but only a bold critic would challenge Gilbert White on what he claims to have seen.

In March 1770 he records, 'Red-wings congregate on trees & whistle inwardly. In their breeding-country they are good songsters: See Fauna Suecia.'[38] He was aware of sub-song, the subdued, hurried singing-in-miniature which in many birds is a springtime precursor to song; he could achieve 'intimate' relationships not only with people but with animals and plants. He helped establish that the stone curlew and the wheatear, as well as the nightjar, were migrants, and that the corn bunting was not. (On the quail, or its alleged movements, he kept an open conscience.)[39] He provided the first accurate description of the 'water rat' or water vole: the animal was an excellent swimmer and diver but – in spite of the confusion of both Ray and Linnaeus – did not have webbed hind feet.[40] His account of the noctule bat, his 'great large bat', was the first English account; and he contributed to the discovery of the lesser whitethroat.[41]

He sent details of his newly discovered '*mus minimus*', the harvest mouse, to Thomas Pennant, in 1767 and 1768, for inclusion in the *British Zoology*. The animal was two inches and a quarter in length, not counting its tail, he writes: two of them weighed down 'just one copper halfpenny' on a scales, 'which is about a third of an ounce avoirdupois'. The spherical nests built by these mice for their young ones were especially noteworthy. Intricately plaited of blades of wheat and 'about the size of a cricket ball', they were securely fixed among the upright corn stalks, well above the ground.[42]

His gradual sorting out of the three 'leaf warblers' was touched on earlier. The work takes on its proper proportion as we notice that while Derham thought they might be three, Ray had made them only two species. The later Montagu (of the harrier) assumed them to be all of one species – one species with a range of songs and calls. The fact was not lost on White that Montagu was an avid collector and gunman; a request from Montagu for bird specimens, after the publication of *Selborne*, was quietly rebuffed. The business of finding one's way among similar species was still highly problematical, and in many cases, White saw, the important clues might be discoverable only by the field observer.

When finally describing the three leaf warblers, he himself had 'specimens of the three sorts' lying in front of him.[43] But the desk-bound naturalist could overlook species, or just as easily, could multiply species, he knew; examined on the study table, and because of age or seasonal plumage, birds of the *same* species could be effectively *separated.* Lack of agreement concerning the names of many quite common creatures, could still create endless problems. A bird Pennant had omitted from his *Zoology*, and which he thought of as a 'willow lark', turned out to be what in White's region was called the 'sedge bird'. White showed that this was Ray's *Passer arundinaceus minor*, the species we know as the sedge warbler. But White himself refers to the same bird on occasion as the 'reed sparrow' – and this confused Daines Barrington. Speaking of the sedge warbler and then the reed bunting, and tactful as always, White explains to Barrington:

> When you talked of keeping a reed sparrow, and giving it seeds, I could not help wondering; because the reed sparrow which I mentioned to you (*passer arundinaceus minor Raii*) is a soft-billed bird; and most probably migrates hence before winter; whereas the bird you kept (*passer toquatus Raii*) abides all the year, and is a thick-billed bird. [44]

4 Densities and distributions

In spite of both tradition and Act of Parliament, White thought that rooks did the farmer more good than harm. This would be decided, he says, if a rook were to be shot, one each week for a year, and the crop opened: 'Tho' this experiment might show that the birds often injure corn & turnips; yet the continual consumption of grubs, & noxious insects would rather preponderate in their favour'.[45] A question concerning the cuckoo – does the cuckoo lay one egg or many eggs – could be answered conclusively, in the same way, if a hen cuckoo were to be shot and dissected during egg-laying time.[46]

White sometimes affects a certain *in*sensitivity, in both *Selborne* and the *Journal*. It can safely be assumed, however, that the dissection of animals *while still alive* was anathema to him. Stephen Hales (1677–1761), a physiologist and experimentalist he knew well during his young manhood, begins one of his papers as follows: 'I tied a middle sized Dog down alive on a table, and having layed bare his windpipe, I cut it asunder just below the *Larynx*, and fixed fast to it the small end of a common fosset; the other end of the fosset had a large bladder tyed to it, which contained 162 cubick inches . . .'[47] White gained much from Hales's experimental conversation, he insisted in later life; but in various fundamentals his usual attitude could not have differed more emphatically from that of the older man.

White was wary of large animals; he was frequently on horseback, and for much of his life must have seen cattle and sheep every day, but he has little to say about horses and farm animals. (His own 'horses', we find, were a succession of steady Galloways.) Conversely, he looked out for and welcomed the diminutive frogs which emerged in great numbers from 'James Knight's ponds', at the bottom of Gracious Street, in May and again in July. The seasonal movements of animals and birds *within* the parish appear from his legible, orderly diaries. In mid-July 1776, he notes; 'Young frogs migrate, & spread around the ponds for more than a furlong: they march about all day long, separating in pursuit of food; & get to the top of the hill, & into the N. field'.[48]

His attitude to earthworms was liberal and well-informed, we said. He observed these animals after dark, on his 'grass plots'. When worms lie out on the turf at night, though they extend their bodies a 'great way', they do not altogether leave their holes, he tells us; they remain sufficiently anchored to be able to withdraw immediately at any sign of danger. Worms can easily find mates even while anchored, being hermaphrodite, and – like the house crickets – they 'engender all the year'. They seem content with whatever foodstuff lies within their reach, and try to draw even fallen leaves into their holes.[49]

White's investigations were not a means of smuggling in something other, or more, than the behaviour of his subjects. He writes in 1779, 'It is now more than forty years that I have paid some attention to the ornithology of this district, without being able to exhaust the subject'.[50] He leaves notes on the fishes to be found in his streams, and on newts ('efts'), hedgehogs, dogs and snakes. What the countryman calls the 'water snake' is none other than the common grass snake, he says, which readily sports in water and perhaps gets some of its food in this element. The grass snake lays 'chains of eggs' in dung or compost in summer, which hatch during the following spring. The young of the viper (adder) are hatched inside the body of the female and emerge as miniature snakes. (He doubts whether female adders 'swallow their young' in moments of emergency, although, oddly enough, he passes on 'salad oil' as a sovereign remedy against the bite of the adder).[51] He finds the freshly sloughed skin of a 'large snake', with the 'scales from the very eyes' intact. Describing this skin, he says; 'If you look through the scales of the snake's eyes from the concave side, viz: as the reptile used them, they lessen objects much.'[52]

He took no inordinate interest in *birds' eggs*, but bird study was his constant, and is his strongest, feature. A true ornithologist can recognise a distant bird by its 'air', we remember, and he describes the distinctive manners of flight of various species and genera.[53] He made only one species – the 'titlark' – of the meadow pipit and tree pipit; they were not finally distinguished as species for another fifty years. But he thought brown owls were responsible for 'all that clamorous hooting'; barn owls, though they screamed horribly and might snore and hiss, did not hoot.[54] This was correct, although the facts regarding the calls of owls were disputed throughout the following century. He was one of the first, if not the first, to refer to the harsh, scolding notes of the nightingale 'once the bird has young ones', and he notes the diligence of parent house martins in taking away the excrement of their young ones from the nest; he refers to the *faecal sacks* which conveniently enclose the droppings of some nestling birds.

In one of the thoughtful observations which did not find a place in *Selborne*, he remarks that those of the summer birds which normally arrive the latest in Britain also *depart* from us the *earliest*: 'this is the case with the hirundo apis, the caprimulgus, & the stoparola' or spotted flycatcher.[55] These birds, we now know, make some of the longest migratory journeys.

In his introduction to the 'hirundine papers', we have another example of his tendency to question an assumption in mid-paragraph without entirely re-writing the paragraph. Here, he praises the 'swallow kind' as particularly useful birds; they keep down the flying insects which would otherwise choke our atmosphere during the summer months. But how does he know the birds have this salutary effect?

No year passes in a village such as his own *without* large numbers of 'swallows' returning as summer visitors. He pauses, therefore, and looks for a control situation. He is aware that some tropical coasts are rendered almost uninhabitable by clouds of flying insects, and he adds, 'it would be worth enquiring whether any species of hirundines is found in those regions'.[56] If no hirundine species live there, he and we will be no further forward; but if 'swallows' are found on those coasts – and he thinks members of the family do nest in Jamaica – the relative absence of flying insects at Selborne may not, in fact, be due materially to this hirundine influence.

He discusses the hirundines quite objectively, at one of his 'levels'. In the four papers dedicated to these species, we are given exact first-hand details concerning their nest-building, laying, incubating and rearing biology. The house martin is closely observed if only because its main nesting material is wet mud:

> As this bird often builds against a perpendicular wall without any projecting ledge under, it requires it's utmost efforts to get the first foundation firmly fixed, so that it may safely carry the superstructure. On this occasion the bird not only clings with it's claws, but partly supports itself by strongly inclining it's tail against the wall, making that a fulcrum; and thus steadied it works and plasters the materials into the face of the brick or stone. But then, that this work may not, while it is soft and green, pull itself down by its own weight, the provident architect has prudence and forebearance enough not to advance her work too fast; but by building only in the morning, and by dedicating the rest of the day to food and amusement, gives it sufficient time to dry and harden. About half an inch seems to be sufficient layer for a day.[57]

In these rightly acclaimed monographs, rather than adding directly to the migration-hibernation debate, he was contributing to the background knowledge which would be indispensable to serious attempts to solve the 'withdrawal' questions. As F. Fraser Darling remarks, 'If you are going to observe an animal well, you must know it well, and this statement is not such a glimpse of the obvious as it appears at first'.[58]

Built up over decades, White's extended behavioural histories were unrivalled in his day. (To be quite accurate, nothing comparable was produced by other workers on vertebrate animals; as will be noticed in Chapter Seven, during the later part of his century the study of insect behaviour could be sustained and detailed.) Over the years, again, he identified 'more than one hundred and twenty' bird species in his locality. Only 252 birds had been identified in the whole of Great Britain by this time, and only 227 were listed by Pennant.[59] Though he did not chase after rare birds, he from time to time encountered true rarities; the black-winged stilts mentioned in *Selborne,* and more notably the wallcreeper described in the *Journal,* were both in this category. White provided the authoritative identification of the wallcreeper, although the record – the earliest of the very few sightings of this bird in Britain accepted in the present-day literature – is summarised as '1792, Stratton Strawless, Norfolk'. The bird was found at Stratton Strawless, and the naturalist first concerned was Robert Marsham (1708-97), but White traced the species. He had been sent drawings of the peculiarly marked wing feathers by Marsham, with whom he exchanged letters for several years towards the end of his life, and from these letters and notes in the *Journal*, his identification was undoubtedly correct.[60]

But his metier was 'manners and modes of life'. The swift's peculiar physiology was not followed up by him, but the same bird's unusual arrival and departure times he regarded as important data. The lives of many quite common creatures – the hirundines were cases in point, despite all the talk and assumption concerning them – had never been methodically examined. He recorded the habits of birds, and very often their numbers or densities; these notes are still of the greatest interest. Early in June 1792, we read with pleasure, 'One fly-catcher builds in the Virginia Creeper, over the garden-door; & one in the vine over the parlour window'. At about the same date, as he was returning in the evening from a neighbouring village, 'Between Newton & us we heard three Fern-owls chattering on the hill; one at the side of the *High-wood*, one at the top of the *Bostal*, & one near the *Hermitage*.'[61] We learn, indeed, that on a late spring or early summer evening, the nightjar, grasshopper warbler, nightingale and cuckoo (or on other occasions, the woodlark), could be heard all at the same time in or from his outer garden.[62] The present-day Selborne resident, though perhaps able to find some, or all, of these birds, would be lucky indeed to equal this record.

The mole cricket was common enough in White's time and region to be troublesome to gardeners, whereas both the mole and (sadly) the field cricket are today rare in Britain. Among summer migrants, the wood warbler and redstart were common birds at Selborne, although the – now uncommon – wood warbler kept, as we would expect, to the canopies of the 'beechen-woods'. In late March or April, a wryneck usually appeared in White's grounds. In 1771, he saw '16 forked-tail kites at once' over the South Downs, and as many as sixty ravens sometimes circled and played over the Selborne hanger. (Today, ravens are virtually unknown in Hampshire, and kites are present in the south of England locally, and only because of re-introduction.) In London, house martins nested even in the sooty regions of Fleet Street and the Strand, and in late summer 'uncountable' myriads of these birds swarmed around the villages of the upper Thames.[63] Choughs, now restricted to the western extremities of the British Isles, nested in the chalk cliffs of the Sussex coast in White's time, and hooded crows, though irregular visitors to Selborne, abounded 'from Andover westwards'. Great bustards were still occasionally reported in Sussex and Hants, and White saw some himself on Salisbury Plain; as he reports, no doubt reliably, 'they resemble fallow deer at a distance'.[64] Corncrakes were sparse but regular summer birds in his parish, and honey buzzards nested in the Hanger from time to time.[65] Corn buntings, though plentiful in more open country, were 'rare birds' at Selborne, and strangely to us, the starling was relatively scarce there,[66] but during the summer months stone curlews – now rare occurrences in any part of southern England – were common and noisy birds on his local downlands.

5 Variable responses

White records crossbills as periodic visitors to a pine grove at Ringmer in Sussex (the home of his aunt, Rebecca Snooke, the original owner of the tortoise); then as now, it seems, there were occasional 'irruptions' of European crossbills.[67] He draws attention to changing populations and distributions. Numbers might be directly affected by human activities, clearly. In the early years of the century, he had reason to believe, the winter flocks of stock doves in his region were very much greater than those he could observe in the 1770s; these birds, though able to take advantage of the increased growing of turnips, could not wholly compensate by this means for the reduced acreage of beech woods, and thus the reduced availability of beech mast, which had come about in his locality in the previous five or six decades.[68] The last of the red deer had been removed from Woolmer Forest by the time he was 'taking down remarks'; and just as regrettably, he says, intensive shooting has wiped out the packs of blackcock, or black game, which previously bred and flourished on this open heathland.[69]

'Soft' winter weather could bring on nest-building prematurely, and weather conditions influenced bird and animal numbers, directly and indirectly. The reduced breeding of house martins during 1776 seemed not to be a result of the cold, wet summer of that year; at all events, during the same summer, the swallows were as prolific as ever.[70] But in hot, dry summers the numbers of partridges in the parish increased greatly, and in the winter months members of the thrush family, and especially the visiting red-wings, to White's surprise, could be thinned very severely by a lengthy period of frost.

Artificial light will keep a captive robin singing later into the night than it would otherwise sing, he notes. Considering the autumn singing of various bird species, he asks, 'Are certain birds induced to sing again because the temperament of autumn resembles that of spring?'[71] Food sources might be directly affected by temperature, and the foods taken by many birds varied with the changing seasons. Not all the soft-billed species leave our shores at the close of summer. During the cold months those remaining live chiefly on 'insects in their *aurelia* state', he reports. Robins and wrens frequent stables and outhouses, and find spiders and flies which have 'laid themselves up', but

> the grand support of the soft-billed birds in winter is that infinite profusion of *aurelia* of the *Lepidoptera ordo*, which is fastened to the twigs of trees and their trunks; to the pales and walls of gardens and buildings; and is found in every cranny and cleft of rock or rubbish, and even in the ground itself.[72]

Wagtails retire to brooks and streams in frosty weather, he can add, for 'there they find the aureliae of water insects'.

In conditions of heavy snow, and when little other food was available, great tits would pull straws lengthwise out of the eves of thatched houses to get at the flies hidden between the straws.[73] Many animals would on occasion take food other than their usual food, out of necessity or merely because the opportunity presented itself. The great apricot tree at The Wakes was more than once robbed of some of its ripe fruit by a dog,[74] and 'willow wrens' among other warblers would eat White's soft fruit in late summer. Ducks had been known to die of eating 'hairy caterpillars', and a tame kite would catch and eat winged ants, 'picking up the female ants, full of eggs, with much satisfaction'.[75] In spite of Ray's appellation, honey buzzards did not eat honey, the 'inquisitive' White was sure; but – from evidence found in their nests – it seemed clear that they or their young ones enjoyed the grubs of wasps.[76]

Swifts and spotted flycatchers, though they normally laid only once during the summer, were both capable of laying and bringing off young ones a second time.[77] Warblers, flycatchers and redstarts hardly do any serious flying while with us in summer but they undoubtedly migrate to distant regions. Wagtails, which if they are not 'flitting' are as a rule low flying birds, will mount high in the air and show a 'great command of wing' when they join together with other small birds in driving off a hawk.[78]

Exchanging notes with Daines Barrington on bird songs, White refers to the variations in key noticeable within the singing or calling of various bird species. A friend with a musical ear has told him that the Selborne cuckoos, for example, may keep one to C, another to D and another to D sharp.[79] Gilbert White was an alert traveller who preferred where possible to travel on horseback (he suffered from coach sickness, but must have appreciated the advantages for the naturalist of the horse as a mode of transport). He knew that the refrains of birds could vary from one region to another. On one of his trips into Sussex, in 1780, he observes, 'Chaffinches sing but in a shorter way than in Hants'.[80] The entry is quoted here in full, and brevity is sometimes an indication of novelty, in White's field notes. More recent naturalists have written dissertations on regional variations in bird song, but in doing so they have been able to call on an accumulation of published material in addition to their own observations. On this occasion, though, White too may have been assisted in advance; a reference to regional variation – or 'provincial dialects' – in bird song can be found in one of Daines Barrington's published papers.[81]

A garden crop might fail quite without apparent cause, and he can refer to 'inexplicable' changes in bird behaviour. In 1791, inexplicably, rooks built in the High Wood at Selborne for the first time, and during the summer of 1791, though young swallows and martins were numerous in the village, not one pair of hirundines nested on or in White's buildings.[82] Alternatively, and in all sorts of circumstances, he describes what seems to him to be problem-solving in animals and birds. The great tits which found flies by pulling straws from the thatch in bitter weather had solved a problem, he thought. (Extraordinary conduct generally occurred in extraordinary conditions, but behaviour might be altogether more flexible than, for instance, the 'reactions to signals' touched on at the beginning of this chapter.) For years he had been aware of an apparently anomalous choice of nesting locations on the part of missel thrushes;[83] missel thrushes, though usually wild and shy, often nested in garden trees, and similarly woodpigeons, which were wary for the rest of the year, might nest near houses and walkways. He knew that, notwithstanding the courage of missel thrushes, their nests were sometimes plundered of eggs and even young ones by magpies. His suggestion – in the last of a sequence of *Journal* entries extending over a period of nineteen years – is that these 'wild, shy' species compromise with human beings temporarily, during the breeding season, as a defence against these predators.[84]

Uncritical anthropomorphism is rare in the work of Gilbert White, but he brought principles to the study of animals and plants which other British writers had brought to the study of human beings. Affect – emotional experience – was expressed in numerous animal responses, he believed; a 'fury' of incubation might possess a broody hen. He saw every reason to, and no reason not to, attribute a degree of intelligence to animals. An example not only of learning but of 'sagacity' appeared, he thought, in a sow of his acquaintance, for of course it is not the case that no larger animals are featured in his writings. For about ten years the sow produced two litters of piglets a year without fail. She had her own 'experienced and artful' way of getting to the boar, on a distant farm, when she was ready to be served. She would proceed to the farm quite unattended, opening all the intervening gates herself, and when her purpose was fulfilled, would make her way home again, by the same route and in the same independent manner.[85]

White occasionally failed to check observationally, and allowed himself to be drawn into *in*cautious claims, for instance in exchanges with Barrington. He was wrong about the 'two vents' with which deer are provided, one at the inner corner of each eye; these vents are not auxiliary breathing inlets; and about the whole alleged business of 'long-billed birds growing fatter in moderate frosts'.[86] He could depart constructively from his outdoor stance. He observed in his own home and in the homes of friends and associates. Domestic and farmyard poultry, the more lowly farm stock, received his careful consideration (as is evident from the second half of D.B. xliii, the letter on 'bird languages'), and the wild birds and animals kept as pets by his neighbours, and for that matter his relatives, are periodically referred to in his notes.

These pets would most of them have been reared by hand, having been found or caught in the wild when very young.[87] He could speak knowledgeably of the indiscriminate appetites of young brown owls, which will eat 'snails, rats, kittens, puppies, magpies, and any kind of carrion or offal', and of the relative difficulty of feeding young barn owls, which will eat only mice.[88] He studied the brown owl kept by John Burbey, the principal village grocer, for many years; it was a 'great washer', and it cast up 'the fur of mice and the feathers of birds' in pellets in the manner of a hawk.[89] In *Selborne* and the journals, we hear of the bullfinches of the village and at various times of a snake, a fox, a squirrel, a bat, a robin, a toad, a buzzard, a raven and even a cuckoo, all of them kept domestically. White did not reject dissection unconditionally, and with the reservations I mentioned earlier, made regular use of it. In the same way, although he kept no pet animals or birds himself, with the exception of the tortoise, which was allowed a great measure of freedom, he frequently accepted the opportunities for exact examination that tame animals provided. Even 'gold and silver fishes' in a bowl showed him behaviour be could not otherwise observe.[90] He avoided menageries – there was one, at least, within easy distance of him when he was staying at South Lambeth – but he might well have visited zoo parks, if there had been such places in his day.[91] He did go to Goodwood Park, in Sussex, to see the moose kept there by the Duke of Richmond. He arrived to find the moose had died, as we shall see, but he asked intelligent questions about why it had died.[92]

Daines Barrington was doing experimental work on bird song with wild birds kept, and in some cases bred, in captivity. The parish naturalist 'knew nothing' of the management of birds in cages, he told Barrington, and he makes no mention of these experiments in either *Selborne* or the *Journal*, but we can assume he was aware of them. (Barrington may well have shown White his birds. But further, the paper in which Barrington gave details of his experiments, and set out his apparently quite remarkable results, was published in the *Philosophical Transactions* and then re-printed as an appendix to Pennant's *Zoology*.[93]) The work bore on the mutability of bird song, or on the parts played respectively by heritable and merely learnt factors in this area of bird life. Bird-trainers played a part in the ornithology of the late eighteenth century – and mutability was in the air in White's time. Despite his dismissive remarks to Barrington, it was a concept underlying many of his own enquiries and 'hints'.

1. Dorton and the Short Lithe, from the churchyard
Photo: the author

2. Part of the Selborne village street
Photo: the author

3. The last of the lime trees planted by White
Photo: the author

4. Priory Farmhouse. Forty pairs of house martins might build under its eaves
Photo: the author

5. The Selborne 'Plestor', or small village green, in White's time
Source: S. H. Grimm, *The Natural History of Selborne* (1789)

Novr: 6: 1767. Selborne near
Alton Hants.

Sir,

It gave me no small satisfaction to find that the Falco which I sent you proved an uncommon one. I must confess I should have been better pleased to have heard that I had sent you a bird that you had never seen before: but that, I find, would be a most difficult thing to do.

I have procured some of the mice mentioned in my former letters, a young one, & a female with young, both of which I have secured in Brandy. From the colour, shape, size, & manner of nesting I seem to make no doubt but that they are non-descript species. They are smaller & more slender than the Mus domesticus vulg: seu minor of Ray: their eyes not so prominent; their ears naked, & standing out above the fur: their tails long, & sparingly covered with hair: their back is redder than that of the Mus domesticus medius of Ray, & more inclinable to the dormouse aspect: their belly is white: a streight line along their sides divides the colours of their back & belly. They never enter houses; are carryed into ricks, & barns in the sheaves; abound in harvest; & (which seems to be their best specific distinction) build their nests between the straws of the standing corn above the ground; & some times in thistles. They breed as many as eight at a litter in a little round nest composed of the blades of grasses.

A Gent: curious in birds, wrote me word that his Servant had shot one last January in that severe weather, which he believed, would puzzle me. I called to see it this summer, not knowing what to expect: but the moment I took it in hand I pronounced it to be the Garrulus Bohemicus, or German silk tail, from the five crimson tags, or points which it has at the ends of five of the short remiges. It cannot I suppose, with any propriety be called an English bird:

6. Letter from White to Thomas Pennant, pages 1 and 2

& yet I see by Ray's Philos: letters that great flocks of
them, feeding on haws, were seen in this kingdom in the winter
of 1685. The mention of haws puts me in mind
that there is a total failure of that wild fruit so conducive
to the support of many of the winged nation. The case is
the same with regard to acorns & beech-mast. For the
same severe weather in the spring which cut off the produce
of the more tender & curious trees, destroyed also that of
the more hardy & common.

Some birds, haunting with the missle-thrushes, &
feeding on the berries of the yew-tree, which answered to the
description of the Merula torquata, were lately seen in
this neighbourhood. I employed some people to procure
me a specimen, but without any success. This species of
Merula belongs properly speaking to the northern & more
mountainous counties; & has nothing to do with Hant.
but by accident.

Quæ: Might not Canary-birds be naturalized to this
Climate, provided their eggs were put in the spring into
the nests of their congeners, such as sparrows, chaf-finches,
&c? before winter perhaps they might be hardened, & able
to shift. About ten years
ago I used to spend some weeks yearly at Sunbury, which
is one of those beautiful villages lying on the Thames near
Hampton-court. In the autumn I could not help being
much amused with those myriads of the swallow kind (I
might perhaps say millions) which used to assemble in
those parts! But what struck me most was, that from ye
time they began to congregate, forsaking the chimneys, &
eeves of the houses, they roosted a nights in the ozier-beds
of the aights of the river. Now this resorting towards the
water at that season of the year seems to give some
countenance to the Northern opinion of their retiring
under water. A Swedish Naturalist is so much perswad-
ed of that fact, that he talks in his Calendar of Flora as

Source: The British Library

familiarly of the Swallows going under water in the begin
ning of Septem.r as he would of his poultry going to roost
before sunset.

An observing Gent: in London writes me word that
he saw a martin on the 23.rd of last Octob.r flying in & out
of it's nest in the Borough. And I myself on the 29.th of
last month (as I was travelling thro' Oxon) saw four or
five swallows hovering round, & settling on the roof of y.e
County hospital of that City. Now is it likely that these
poor little birds (which perhaps have not been hatched
many weeks) should at that very late season of the year,
& from so midland a county attempt a voyage to Goree,
* or Senegal almost as far as the Æquator? I acqui
esce entirely in y.r opinion, that tho' most of the Swallow kind
do migrate; yet that some do stay behind, & hide with us
during the winter.

As to the short-winged, soft billed birds, which come
trooping in such numbers in the spring, I am at a
loss even what to suspect about them. I watched
them narrowly this year, & saw them abound till
about Michaelmas, when they disappeared. Subsist
they cannot openly among us, & yet elude the eyes of the inquisi
tive: & as to their hiding, no man pretends to have found
any of them in a torpid state in the winter. But with regard
to their migration, what difficulties attend that supposition,
that such weak, & bad flyers should be able to traverse vast
seas, & continents in order to attain milder climes amidst
the regions of Africa?

As to what I sent you word about the lampetra cæca
being found in our streams, I desire to retract it; having
Reason to suppose that those I saw were the lampetra
parva, & fluviatilis of Ray. Nothing would vex me more
than to find I had misled by my inaccuracies a Gentleman
who intends to favour the world with his researches into
natural History.

* See Adamson's voyage to Senegal.

7. Letter from White to Thomas Pennant, pages 3 and 4

Loches abound in the stream that runs by Ambersbury in Wilts: & that rare fish the rudd or finscale I have seen procured from the Charwell near Oxford.

Begging the continuance of yr most agreeable correspondence, I conclude with great esteem,

Your most obedient Servant,
Gil: White.

P. S. What parts of England does the goss-hawk frequent?

To
Thomas Pennant Esq
at Downing
Flintshire
North Wales.
A single sheet.

Novr 4th 1771. Mr White saw 3 swallows on the wing near Beachy head. probably says he the warm sun had drawn them from their Hibernacula in those chalk cliffs which front the south & hang over the seas which circumstance must contribute to keep the air there in a temperate state.

Source: The British Library

8. Brown owl and nightjar (fern owl or goat sucker)
Source: Thomas Pennant, *British Zoology* (1761), vol. 2

Chapter 6

INSTINCT AND INITIATIVE

1 Insightful response and automatic reaction

Gilbert White's base was the second half of the eighteenth century, but two hundred years on, he can seem to have been a time-traveller: he was a representative of his time who seems also to invade our own time.

In one of the *Journal* entries on the rookery at Faringdon, he refers to courtship feeding among pairs of rooks:

> As soon as rooks have finished their nests, & before they lay, the cocks begin to feed the hens, who receive their bounty with a fondling tremulous voice & fluttering wings, & all the little blandishments that are expressed by the young while in a helpless state. This gallant deportment of the males is continued thro' the whole season of incubation.[1]

As far as I am aware, White was alone among eighteenth century naturalists in referring to courtship feeding; and his description of this intimate, ritualised conduct immediately recalls David Lack's introduction of the same topic in his study, *The Life of the Robin* (1943). According to Lack,

> In courtship-feeding the hen robin hops about uttering a loud call, then, as the cock approaches with food, she partly lowers her wings and quivers with excitement, while the call changes to a rapidly repeated note as the cock finally feeds her. The hen's attitudes and calls are indistinguishable from those of a young robin being fed by the parents.[2]

Regarding courtship ritual extending beyond coition and egg-laying, Lack remarks that Julian Huxley was the first to describe such ritual prolonged in this way. He notices the explanation of this behaviour proposed by Huxley; it helps ensure that the 'bonding' between the two birds will continue not only during incubation but throughout the period of brooding and feeding the young ones. Lack had overlooked Gilbert White on courtship-feeding; and although White offers no interpretation of this behaviour, a further *Journal* entry can be added to that concerning the rooks. In May 1780, White refers to courtship-feeding among robins. The gallant male robin, he notes, 'feeds his hen as they hop about on the walks, who receives his bounty with great pleasure, shivering with her wings, & expressing much complacency'.[3]

David Lack was an 'ethological' naturalist; he was part, that is, of a broad movement among zoologists and animal psychologists which took on substantial identity during the 1930s and has steadily increased its influence from that time to the present day. In its early stages, ethology represented a reaction against both academic zoology and an animal psychology developed largely in the laboratory.[4] The ethologists, though they comprised several schools from the beginning, were interested in the spontaneous doings of animals. The laboratory work they rejected was more neurological than psychological, and was restricted by a preoccupation with reflexes; in its studies of rats learning to run mazes or pigeons dealing with puzzle boxes, it minimised spontaneous behaviour and prejudged the behaviour of animals in their usual surroundings.[5] Moving out of the laboratory and into the countryside – in Konrad Lorenz's case, concentrating very often on what he called his 'family retainers', animals such as his geese and jackdaws which were semi-domesticated but which led their own lives – the ethologists, by contrast, allowed instinct a role which was then unfashionable. The responses of total – intact – animals were in many cases innate, they insisted, and these responses were at least as fundamental to the lives of the animals concerned as even modified reflexes. Given the chance, animals evidenced both fixed habits and general preferences; whether or not they contemplated goals, they would work and learn with more alacrity in some areas of possible experience and encounter than in others. In the words of Niko Tinbergen, a stereotyped response might be preceded by what looked like a purposive 'searching' on the animal's part.[6]

The ethologists allowed animals to find their own problems, or at least their own methods of problem-solving. In this they had a forerunner in Gilbert White. They had a forerunner in White, too, in their understanding of instinct – as Lack seems to agree. If Lack had not read the *Naturalist's Journal*, he was familiar with the *Natural History of Selborne*; in his book on the robin he uses an extract from White to introduce his chapter on instinct:

> Philosophers have defined *instinct* to be that secret influence by which every species is impelled naturally to pursue, at all times, the same way or track, without any teaching or example; whereas *reason*, without instruction, would often vary and do that by many methods which instinct effects by one alone. Now this must be taken in a qualified sense: for there are instances in which instinct does vary and conform to the circumstances of place and convenience.[7]

White can refer to instinct as a 'wonderful, limited' provision, but he sees that in a sense an animal's being impelled to behave in a given manner without teaching or example may raise it 'far above' reason. Instinctively activated, a domestic hen will fly in the face of a marauding dog or sow – animals she would normally shy away from – in defence of her young ones, and as an 'observer at Gibraltar' has informed him, a blue thrush will dart with similar courage at a kestrel or sparrow hawk which approaches its nest.[8] In circumstances such as these, automatic reactions are more reliable, and faster, than would be a thought out response. Ready-made and automatic responses are indispensable to young animals, similarly, creatures which have not had time to learn about the dangers which may suddenly threaten them. A young game bird crouches motionless, a domestic chick seizes a fly but avoids a wasp, instantly and as a matter of innate preparedness. An infant viper evidences fully formed aggressive responses, rearing up and gaping as if ready to strike, literally from the moment of its birth. White could add this further example because he had been present when a female viper 'bloated with young' was killed in his aunt's garden at Ringmer. He had opened up the newly dead reptile, and fifteen vipers the size of earthworms had emerged from the body of their parent already exhibiting this 'true viper spirit'.[9]

To a thinking person, nothing is more wonderful than

> that early instinct which impresses young animals with the notion of the situation of their natural weapons, and of using them properly in their own defence, even before those weapons subsist or are formed. Thus a young cock will spar at his adversary before his spurs are grown; and a calf or lamb will push with their heads before their horns are sprouted.[10]

Clearly, there are also disadvantages to innate and automatic responding. A creature depending only on habits such as these may on occasion make a fool of itself – and far from saving its life, may forfeit it, as a result. The instinctive crouching of game birds protects them against stooping hawks, but the same habit renders them especially vulnerable to another danger, that of human beings with guns.[11] 'Instinct' has various connotations in White's notebooks and essays; it usually means *inborn fixed reaction* but it may include *inborn propensity and need*; 'instinct' overlaps with 'motive' in his writings.[12] But repetition may be profitless and even dangerous. A farmyard hen possessed by the need to incubate will sit on and 'try to hatch' a single stone, and if a female turkey's eggs are taken away, she may sit persistently on nothing at all.[13] Perhaps most interestingly, the hirundines could provide him with examples of creatures 'misled by instinct'. House martins which have attempted to build against an exposed window, and whose partly finished nest is washed away each time it rains may struggle doggedly on, never completing a nest. They may even return for several years running to the same unsuitable site, and to a 'one track' repetition of the same fruitless activity.[14]

As well as inborn behaviour, White was aware of behaviour which seemed to *depend 'entirely' on* teaching, or better, training. He knew of Daines Barrington's work on the education of song-birds, we said. Barrington had demonstrated that various birds, if brought up from birth or from a very early age with birds of another species, will call and sing exactly as do these other birds. Linnets which had learned to sing under skylarks or woodlarks 'adhered entirely' to the songs of their instructors, and a robin trained under a 'skylark-linnet', a linnet, that is, which had itself been reared and influenced by a skylark, sang in its turn in the manner and style of a skylark. A house sparrow, even, reared near a singing goldfinch, adopted part of the goldfinch's song. Song-birds which grow up under no instructors, alternatively, do not sing at all, Barrington had found; they make only elementary whistling and chirruping sounds. As the educationalist concluded, then: 'Notes in birds are no more innate, than language is in man, and depend entirely upon the master under which they are bred, as far as their organs will enable them to imitate the sounds which they have frequent opportunities of hearing.'[15]

Barrington was an indoor naturalist, and when he asked for live specimens, White told him, 'I am no birdcatcher'. But the question of the extent to which a creature could adjust to unfamiliar conditions was one White himself often returned to. He asks in a letter to Pennant, 'Might not Canary birds be naturalized to this climate, provided their eggs were put, in the spring, into the nests of some of their congeners, as goldfinches, green-finches, &c? Before winter perhaps they might be hardened, and able to shift for themselves.'[16] He took some field crickets from the Short Lythe and liberated them on his terrace, boring holes to help them settle in. This attempt was only half-hearted and, like several similar 'experiments', it was not successful. The crickets wandered away by degrees, singing at a greater distance from the garden every morning, and returning, as it seemed, to the place they had been taken from.[17] But then, White's predominant outlook differed from that of Barrington. In the letter to Pennant, he was wondering what would happen to canaries hatched and reared by British goldfinches or greenfinches in the wild. His own work on bird songs and calls was, and could only have been, conducted in the 'outdoor' habitats of the birds in question. The songs and calls 'served purposes' within highly complex social and physical environments, and learning as he generally considered it took place within behavioural fields the creatures in question needed to operate in. The famished great tits found insects in the thatch, the sow who needed the boar applied herself successfully to the problem of the intervening gates. Even Barrington's 'entirely factitious' young song-birds, he might have suggested, were proof of a readiness on their part – a readiness to be taught to sing.

In the wild, he might have pointed out also, birds may hear the refrains of numerous bird species while learning to sing, and yet distinct songs are perpetuated within given species. (It is now quite clear that while some birds, such as the starling and the sedge warbler, are eager mimics even in their wild state, others – for instance the blackbird and the chiff-chaff – show little tendency to learn songs other than those normally associated with their species even when attempts are made to train them artificially. As W H Thorpe was to suggest, the chaffinch is typical of many birds, in the way its singing develops: the rudiments of the song are inborn, but the stylistic details, including the precise tone and rhythm, have to be learned by young chaffinches from the mature birds.[18]) White would not have moved from results relating to certain song-birds in cages to dogmatic statements about birds in general, in the manner of Barrington; and his 'outdoor' orientation worked against his dividing the behaviour of an animal into quite independent components. He was interested in all an animal or bird did. That a domestic hen has chickens is apparent not only from her vocal expression but from her behaviour at large, he tells us; her attitude to the other hens and even her walk expresses this new situation, and she is capable of greater bravery and greater cleverness – 'invention' – than a hen would ordinarily exhibit.[19]

As White thought of it, 'intelligent' learning was part of a larger pattern. He notes during one harvest time:

> Many creatures are endowed with a ready discernment to see what will turn to their own advantage & emolument; & often discover more sagacity than could be expected. Thus Benham's poultry watch for waggons loaded with wheat, & running after them pick up a number of grains which are shaken from the sheaves by the agitation of the carriages.[20]

Benham's poultry showed 'discernment', or heightened discernment, in certain connections; they were quick-witted, in this example, in matters concerning their own feeding. He refers to animals which have dealt successfully with obstacles to the fulfilment of their needs. As an 'intimate' observer, he was aware of differences between creatures of the same species, and (unlike William Borlase, who admittedly was referring mainly to farm animals) he saw that these differences might be vital to the 'lives' the creatures led. Encountering problems in regard to their nest-building, house martins might behave quite differently from the house martins which toiled on unavailingly in their exposed window. In nesting under the eaves of one of the houses near White's own in the village, birds of this kind had to cope with a protruding joist, which left them only a narrow space to work in. As a result they built 'flattened' or 'tunnel-like' nests. When first meeting with this obstacle, this is to say, they did not keep, or abortively try to keep, to the hemispheric section which is the house martin's usual nest form.

Chaffinches and wrens nesting on the outskirts of London, he found, had adapted their building techniques to the materials available there, or to the absence in those localities of various of the nesting materials normally used by members of their species. The country observer – the Selborne observer – might assume the nesting styles of both chaffinches and wrens to be unalterably inherited, so predictable are they in his experience; but when necessary some birds of these species can depart from the 'universal' pattern. The suburban chaffinch managed without the usual lichens and gossamer, we are told, and in the same situation, the suburban wren used only 'straws and dry grasses'.[21]

At Chilgrove, the village near Chichester which White visited annually, 'many jackdaws' had for years nested in parts of a large rabbit warren. Puffins nest in burrows, he remarks to Pennant, but 'I should never have suspected the daws of building in holes on the flat ground'. In his next letter, he adds, 'As to the peculiarity of jackdaws building ... under the ground in rabbit burrows, you have, in part, hit upon the reason; for in reality, there are hardly any towers or steeples in all this country'.[22]

Under pressure of parental loyalty, some spotted flycatchers which were nesting in a vine on White's house exhibited a more startling adaptability. Building during a 'shady time', they had chosen a position such that when their young were hatched, and when hot weather came on, the young ones were in danger of being quite destroyed by the heat of the sun reflected from a wall. The expedient the parents resorted to was that of making themselves into avian umbrellas and fans. They hovered over the nest and its contents 'all the hotter hours', with wings expanded and mouths gaping for breath, and in this way screened and cooled the helpless nestlings.[23] It was not a stratagem they had 'learned from experience'; the difficulty they were dealing with was novel; and for the same reason – and though we might not argue in quite the same way today – this could hardly have been an inborn reaction. Spurred on by a felt involvement with the endangered young ones, White implies, the adult birds had reached an ingenious and plucky solution.

We can see evidence of a 'wonderful spirit of sociality' in many animal species, he says. If he was aware of individual differences, he was aware too of a gregarious need in many of the animals he studied. A horse which has been shut up alone may break out of its stable, merely to be with other horses. Similarly, cattle will not fatten by themselves; they 'will neglect the finest pasture that is not recommended by society'.[24] The hirundines congregated and 'conferred', prior to their departure, and the finches which nested in strictly separated pairs in spring and summer became, as it were, quite different birds as they gathered into their winter flocks.

The need to be with companions no doubt explained various cases of anomalous attachment. The naturalist knew of a doe which had grown up from an early age with a 'dairy of cows'. The doe went out to the field and then returned to the yard with the cows each day, and the cows, on their side, treated the doe as one of themselves; they would belligerently drive off a dog if it tried to molest her. In the absence of groups, in the absence of all other companions, a horse and a solitary farmyard hen formed a close friendship. The regard between the two was such that 'The foul would approach the quadruped with notes of complacency, rubbing herself gently against his legs: while the horse would look down with satisfaction, and move with the greatest caution and circumspection, lest he should trample on his diminutive companion.'[25]

Broody hens without chickens will foster ducklings, and a nursing cat, White tells us, adopted a new-born hare, a leveret, in place of her own kittens when these had been taken away. (In a similar incident noted in the *Journal*, a cat was offered a family of baby squirrels, and fostered them.) The cat suckled and brought up the leveret, he says, even though on other occasions she might have been pleased to catch and eat just such a tender morsel. From his description, it is clear that the leveret, in its turn, had adopted the cat.[26] The leveret had imprinted the cat, to use ethological language, and the detail is added that when the cat eventually returned to the house, the leveret was following behind her.[27] Because of events such as these, the Selborne naturalist continues, even stories of human children brought up by wolves and other wild carnivores cannot be uncritically dismissed.

Animals which usually keep to a given life-style are or were not, merely on this account, incapable of other life-styles; the instances of exceptional parental behaviour are cited by White not as oddities but, on the contrary, as throwing light on the usual parental behaviour of hens, cows or cats. White, however, though he attributes insight and even intention to the 'brute creation', does not do so gratuitously. Whatever we are to think of the 'loyal' spotted flycatchers, the initial response of the bereaved cat which was offered the leveret, as he describes it, has something in common with the automatic crouching of the threatened game bird. More particularly, it recalls the fixed and predictable responses to 'emblematic' postures, colours or sounds touched on in the previous chapter. The cat was ready for the maternal role, and having just lost her own newborn kittens, would have been beset by a 'tender desiderium'; but in White's opinion, other more specific circumstances helped precipitate her involvement with the leveret. She was distended with milk, for example, and 'experienced relief' when her teats were sucked. She no doubt became used to the presence of the leveret, but White's eventual suggestion here is 'stranger' than this. The cat was or became 'as much delighted with this foundling as if it had been her real offspring'.[28] In her condition, she did not need the encouragement kittens and only kittens could provide. The sucking itself, among other special stimuli, elicited the maternal concern and conduct; and given these stimuli – and bereaved as she was – the cat would have been unable *not* to respond to the young hare, or some other comparable creature, in this maternal manner.

This triggering of involuntary responses seems similar to the 'releasing' recognised by the ethologists of our own time. (Releasing may result in surprising relationships.[29]) And White can relate even this to a broadly predisposed behaving creature, or to a view which, while it was accepted by mid-twentieth century ethologists, is now regarded much more critically. Two 'great motives' underlie and regulate all animal behaviour, he says. The two are 'love' and 'food'. Of these, 'the former incites animals to perpetuate their kind, and the latter induces them to preserve individuals'.[30]

Much animal behaviour can be readily assimilated to one or other of these 'great motives', White thinks; but how can, for instance, the winter flocking of finches be thus accounted for? Mating is not a winter activity – and when it does take place these species are established in territories. As to 'food', or the self-preservation motive; surely this congregating in a time of austerity must make it more difficult, if anything, for the individuals to find sufficient food. One view White considers is that the finches form flocks 'from the helplessness of their state' in winter conditions. They crowd together for protection and perhaps warmth, and in addition feel ('seem to themselves to be') more secure in these large numbers. Much as human beings might, they *gain encouragement* from the presence of their fellows. The survival of both individuals and species is furthered indirectly, then, by this winter flocking.[31]

'Food' and 'love' appear in more than one role, in White's reporting and tentative discussion. They sometimes show us the functions of animal behaviour, rather than its causes. As was noted in the previous chapter, functions, the 'purposes served' by given habits, were among White's continuing interests. (Maternal solicitude was perhaps the more efficient, for being involuntarily triggered.) But as these motives are presented first, they are felt needs, and White, we find, can class several others with the big 'two'. Maternal need can itself be 'the strongest of motives', he says; and 'sociality', or gregarious predisposition, can operate quite independently of sexual appetite.[32] The principal 'motives' are few in number, this is the claim here. They direct activity and at the same time allow animals, or some animals, a potential latitude.

A habit might be permanently readjusted, White saw. He refers to factitious but *now commonly observable* styles and habits. He emphasises the virtually continuous 'conversation' of animals with other animals, and touches, frequently if tangentially, on the notion of an adopted adaptation which has spread through a population – for instance a suburban population – or even a species.

2 Emotion, education and tradition

He refers to spontaneous behaviour but he believes also that no animal or plant is altogether self-sufficient. He records the *shiverings*, *struttings*, *hoppings*, '*swoonings*' and *peculiar cries* he thinks of as observable signs of 'passionate' – emotional – arousal. A great many actions and reactions, and all inter-personal and social relationships, are emotional in one of their aspects; feelings may cement friendships between quite diverse creatures, as with the horse and the hen, and feelings united the finches in their winter flocks. He is inclined to put maternal animals on a pedestal; he regards with horror any 'perversion' of maternal feeling.[33] (Had he, one wonders, been denied his share of maternal affection as a child?) He would have been less ready to accept the 'automatic' and up to a point indiscriminate operation of maternal response, we can say, had he not been able to consider the responses of cats which were nurturing their own kittens and the case of the cat and the leveret comparatively, and had he not been able, after this, to compare various examples of anomalous maternal attachment, in various species, with each other.

His analytical examination in no way detracts from the reality of maternal (or parental) affection.[34] This affection may in due course be replaced by 'jealousy', we saw, but the jealous behaviour does not appear until the youngsters have attained to relative independence. For a vital period of days, weeks or even months, the fledged juveniles of many bird species are supervised and guided by their parents.

Swallows and martins feed their newly emerged fledglings, first as 'perchers' and then as 'flyers', White tells us, tiding the young birds over until they have adjusted to the work of hunting for food.[35] Adult swifts seem to direct their fledged young in the acquiring of greater areal skills; on July evenings, 'Swifts get together in a large party, & course round the environs of the church, as if teaching their broods the art of flying'.[36] In many bird species, parents show their young ones where to find suitable food. The young of stone curlews run almost as soon as they are hatched, and the parent birds, if they do not feed the chicks, seem to 'lead them about at the time of feeding'.[37] Rooks, which are early breeders, bring out their fledged young ones for the chafers in May, and whitethroats, which nest well away from human habitations, 'bring their broods into gardens and orchards' in July and August for the first soft fruit.[38] In early September, the stone curlews which flew over White's garden at night, going to and from Dorton to feed, were accompanied by fledged juveniles. This dctail was adduced by the careful observer, and even more careful reporter, as he noticed and on later occasions confirmed that a 'piping, wailing' call had by then been added to the notes of the adult birds.[39]

White's awareness of 'natural economies' was demonstrated earlier. In the Selborne parish, the planting or removal of trees resulted, and the destroying of all earthworms 'would have resulted', in consequences far exceeding their obvious consequences. Similarly, and though he sometimes makes no clear distinction between original problem solving and a problem-solving tradition, he assumes that the one may give rise to the other. Summer migrants might use the same nesting sites over periods of many years; young animals were not merely inactive recipients, but regarding the swifts which had 'always' nested under the thatched roofs of certain of the Selborne cottages, it was likely that 'some of the same family still returned to the same place'.[40] The use of rabbit burrows by jackdaws at Chilgrove was 'exceptional' but was clearly an established traditional expedient. The effects of social and cultural norms on human lives were well appreciated by White; in a superstitious community, he says, where no 'liberal' upbringing is at work to counteract this influence, irrational fears may be so implanted in young children as to be virtually ineradicable in the adults they become.[41] In much the same way, he seems to have thought, local solutions may be perpetuated within non-human families, groups or populations.

He suggests as well that local solutions may in time come to be genetically passed on; innovative or merely acquired behaviour may sooner or later affect the directly heritable make-up of a population. Changed heritable adaptation, or White's circling and yet persistent approach to this wide topic, will require a chapter to itself; although it can be mentioned here that on the question of the *process by which* an acquired habit might get into the genetic legacy – to us, in retrospect, it seems the vital question – he offers no opinion.

Cautious on various topics, he refers to behavioural study itself quite unequivocally; it is often 'the best part' of natural history.[42] In this chapter, I have compared his attitude to that of the twentieth century 'ethologists'. Several of their immediate predecessors, in various countries, are equally relevant. Edmund Selous, the brilliant early twentieth century fieldman and observer, dwells on the importance of signals and symbols. He insists too that 'the "brute beast" is a more intelligent, more emotional, more affectionate and generally fuller-feeling being than he has as yet been acknowledged to be'.[43]

A scientist's work is validated or invalidated as it is found to be more or less reliable; 'found', that is, by other researchers who repeat the work concerned in as near as possible the field or laboratory conditions in which it was originally carried out.[44] Most of the questions White had begun to ask and answer were not seriously taken up again until the time of the much later naturalists just referred to. His 'observing and comparing' has been repeated, and his results have been largely substantiated, but only after a century or two of hiatus. I shall imply that a knowledge of twentieth, and now twenty first, century 'ethology' and 'ecology' – and the use within these of comparative, analytical techniques – can help us assess White 'as naturalist and scientist' more appropriately. Such a view can be contentious; it is sometimes insisted today that the science of one era cannot be usefully compared with the science of another. However, my own suggestion is not that events which had still to take place, in particular twentieth century events, somehow *explain the occurrence* of White's – eighteenth century – work. In this book scientific and historical discussions are being conducted side by side. The two sorts of discussions, or questions, are different; they need not obstruct each other, even if they do in some respects affect each other. Because of the seeming 'modernity' of the Selborne naturalist, the 'historical' question, the question of the influences, cultural, economic and so on, which affected his work, takes on a particular urgency. It will be returned to in, notably, Chapters 8, 9 and 11. We cannot take White's exceptional talents away from him, but in this historical and explanatory connection we ourselves have to add the rider, 'this cause but not this cause alone'.

But we have by no means finished with the observational records he leaves us. (For the present I am assuming that we can take these at what is to us their face value.) In the next chapter I want to look at his notes on the 'specific' and locally adapted manners of plants and insects.

Chapter 7

BEHAVIOUR: PLANTS AND INSECTS

1 Sexuality in plants

The behaviour contributing to the perpetuation of an animal species, or which could be brought under the heading 'love', included sexual behaviour, of course. White did not dwell on sexuality, but he referred to it without prurience and without embarrassment (to the displeasure of some of his Victorian editors). He described the coupling of swifts in the air, and if he was puzzled as to 'how the female toad was impregnated', he was one of the first, if not the first, to recognise the hermaphroditism of earthworms.[1] He willingly accepted the importance of sexuality in the lives of plants, although this appears more from the journals than from *Selborne*. The botanist too, if he was to make sense of his subject, had to allow for love – or the indispensable contributions of both males and females.

If the idea of plant sexuality was not new, it was still being got used to, when White first turned to botany. Sexual reproduction in plants had been discussed since the end of the seventeenth century, and the 'sexual system' of classification had been proposed by Linnaeus in 1735, but the major Linnaean botanical works had appeared only in 1753 and 1754. White began using the sexual system – and the Linnaean binomials – from 1767 onwards. (In the later years of the *Garden Calendar*, he can be found using Ray or Linnaeus or both. He acquired Hudson's *Flora Anglica*, which was Linnaean with modifications, in 1765.) But he saw that the knowledge of plant sexuality gradually progressing in his day had important applications in addition to those concerning identifying and taxonomy. With plants as with animals, sexuality was the key to much of their observable behaviour. In a case he himself considered and discussed, sexuality, or its repression, was perhaps the key to their unsatisfactory behaviour.

The case concerned the hop growing which was important in the Selborne parish. Hop growing was a risky undertaking, because, 'more than any other farm crop', hops were subject to 'unaccountable blights and failures'. These failures could not be related immediately to the weather, for instance, or to insect pests. White came to believe that the farmers were themselves to some extent responsible for them; they were ignorant as to the total needs of the growing hop plants. The hop is dioecious, and the practice was to plant or retain in the hop gardens only the *pistillate*, or female, hops. The reason, no doubt, was that the *staminate* or male plants did not produce the large, papery hop-heads used in brewing, and could be regarded, therefore, as taking up space and nourishment unnecessarily. The hop harvests were often successful; but White, walking about their hop gardens with the growers, put it to them that perhaps they had got things the wrong way round. Perhaps some of the staminate hops were escaping notice, and were one cause of the *good* harvests, and perhaps the assiduous removal of all staminate plants, where it was achieved, was *causing the bad* harvests. Male plants, even though they do not bear good hops, may be necessary to the production of good hops:[2]

> hence perhaps it might be proper, tho' not practiced, to leave purposely some male hop plants in every garden, that their farina might impregnate the blossoms. The female plants without their male attendants are not in their natural state: hence we may suppose the frequent failure of crop so incident in hop grounds.[3]

Just how much notice the growers took of his suggestion we do not know, although by 1791 John Hale, the farmer-butcher in the village, seems to have been acting on it. White records in August of that year, 'In Mr Hales's hopgarden near Dell are several hills containing male plants, which now shed their farina: the female plants begin to blow.'[4] The question was debated in the nineteenth century, and today seems by no means settled.[5]

Not all farina (or pollen) is wafted on the air, and White knew that a reciprocal benefit accrues as bees visit flowers. Bees are 'much the best setters' of cucumbers, for example. He recommends putting a little honey on the male and female flowers to tempt the bees into the cucumber frames,[6] and he sends some 'male cucumber-blossoms in a box' over to Newton Valence, to set some cucumbers in flower in the Newton vicarage frames.[7] The subtleties of cross-fertilisation were only beginning to be understood. If he sometimes implies that no plant can bear fruit unless it has been fertilised,[8] this is in keeping with his recognition of the precariousness of successful plant reproduction. Wheat is bisexual, but if it is to fruit or seed well, it must first fully blossom, he notes: 'Windy, wet, cold solstices are never favourable to wheat, because they interrupt the bloom, & shake it off before it has performed it's function'.[9]

Yew trees or their flowers carry all male or all female parts, and he thoughtfully records the genders of five isolated churchyard yews; two are female trees and the other three, including the Selborne yew, are all male trees.[10] Sexuality in mosses could be ascertained, but the flowerless 'cryptogamia', as their name implied, were a propagation mystery.[11] White was a bachelor, but he shows no willingness to impose celibacy on other, less trammelled creatures.

2 The observational botanist

He took to wild as well as garden plants with enthusiasm towards the end of the *Garden Calendar* period. Increasingly from the mid-1760s, we find him using the 'sexual system', the first, and in the eighteenth century the most influential, Linnaean method of identifying and classifying. He records, in August 1767:

> Discovered the yellow centory, Centaurium luteum perfoliatum of Ray, in plenty up the sides of the steep cart-way in the King's field beyond Tull's. This is a very vague plant for ascertaining according to the sexual system. Linn: makes it a gentian, & places it among the pentandrias: but it has commonly seven stamina. Hudson makes a new Genus of it (Blackstonia) unknown to Linn: placing it as an 8 andria digynia. It is best known by it's boat-like, very perfoliated leaves.[12]

During 1766 he compiled his *Calendar of Flora*, the third (chronologically, the second) of his diary/notebooks, a one-year record of flowering plants, cultivated and wild, in his locality. The plants are classified and are listed in the order in which they appeared or flowered. This calendar, though produced only a year after he had bought his Hudson, covers some four hundred wild plant species, and according to Walter Johnson, altogether, at one time and another, White identified at least four hundred and forty wild plants in his parish.[13]

Identifying and classifying were indispensable to botanical collecting. The archetypal status of these activities for the amateur naturalist is clear from the view stated by Benjamin Stillingfleet: 'He who goes farther than this is not barely a naturalist, but something more, viz. a physician, a chymist, a farmer, a gardener, &c. And he who cannot go thus far to a certain degree, does not deserve the name of naturalist'.[14] White had extended his knowledge of contemporary, continental taxonomy in the course of tutoring John White. He agrees that nature would be 'a pathless wilderness' without an accepted systematics, and he could express great respect for Linnaeus (1707–78). He saw the dangers of *mere* taxonomy as an objective, however, dangers which were evident from the sexual system itself. This 'system' depended on two facts: (1) a great many plant species have 'stamens and pistils', whether or not in a given species these are found together in the same flower or on the same plant, and (2) these sexual parts occur in greatly varying numbers and ratios, depending on the species. The numbers and arrangements of the stamens and pistils, or in practice stigmas, in any instance could significantly assist with plant identification; but equally, by means of the sexual system, plants having *only* the numbers and arrangements of their stamens and pistils in common could be gathered into the same taxonomical category. (The tiny Lapland 'gentian' has six stamens and one stigma in the same flower head, or as Linnaeus puts it, strikingly enough, six husbands and one wife in the same bed. It becomes bracketed, therefore, with the wallflower, which, though it has the same combination of male and female parts, belongs to a quite different 'natural genus'.)[15] White appreciates the 'ingenuity and precision' of the sexual system, but the botanist, he is sure, must be more than a 'maker of lists'.[16] White would go further than Linnaeus towards recognising the mutability of species, but he nowhere doubts the reality of species or genera. With plants as with animals, he nowhere doubts the reality of the biological relationships between what he calls 'congenerous' species.

Various of White's statements on identifying and classifying occur in one of the 'letters to Barrington' (DB xl). His alternative – for the botanist who is not to be a mere maker of lists – has sometimes been found in the same letter. A selective approach to plants has been requisite to civilisation itself, he says there, and 'instead of examining the minute distinctions of every various species of each obscure genus', the botanist should try to make himself acquainted with 'those that are useful'. He should ally himself with, most notably, the gardener, the husbandman and the physician.

The words recall Stillingfleet's remark, and the production of White's *Calendar of Flora* was without doubt influenced by Stillingfleet. White acquired a copy of the second edition of Stillingfleet's *Miscellaneous tracts* (1762), at some time during 1764-5. In this work Stillingfleet had included two plant 'calendars', one compiled by himself in Norfolk, the other by one of Linnaeus's colleagues in central Sweden, both for the year 1755. Calendars such as these could have a practical and more especially a horticultural use, Stillingfleet was insisting. By learning the order in which plants emerge or flower in his region – and by paying 'some attention' to the arrival of migratory birds and the conditions in which hibernating species become active – the gardener or farmer could provide himself with a guide to when, in any year, he should do his own sowing and planting out.[17] Stillingfleet thought that something might be gained if links were found eventually between guides built up in this way in various European countries, but he was principally interested in the local sequences themselves.

Nature diaries became a feature of the 'popular nature study', and in the preface to his printed diary blank, Daines Barrington cites Stillingfleet as an example other amateur naturalists should follow. White, however, though he produced a calendar of flora, makes little mention of Stillingfleet's hopes. (Even where White started as a follower, he soon began questioning and testing.) At Selborne, he knew, though the 'natural indicators' would often be reliable, the usual order of events might quite break down in a given year. The summer birds could fail to reappear in the 'right' local weather conditions – and could turn up instead in the 'wrong' local weather. Trees or flowers could be brought forward by a brief warm spell in what was otherwise a cold and late spring. Virgil had recommended the study of local indicators;[18] but the farmer or horticulturist who depended on aids of this kind uncritically in southern Britain, White could see, would soon meet with disaster.

In the same way, he did not implement the conventional utilitarian plan he appears to accept in the 'letter to Barrington' just referred to. In practice, the subordination of botany to agriculture, medicine and so on was not his alternative to mere classification. His alternative was the study of the lives and relationships of plants under various local conditions. An enhanced understanding of the 'laws of vegetation' will be useful: along with an outdoor and behavioural zoology, it will contribute to the 'good of mankind'. But the usefulness will be the result of the researcher's having accepted scientific aims, not of his having restricted himself to socially approved topics.

There were plants for which White felt a particular sympathy, such as the whisp of ivy-leafed toadflax which established itself and then made annual advances on one of his house walls. But the botanist who is *ecological* (for want of a better term), *analytical* and *predictive* will be of use to others. Such a botanist may be able to show some of the consequences to be expected from the introduction or removal of a given plant. During the eighteenth century, as the beech woods in White's part of Hampshire had been reduced, the enormous numbers of stock doves arriving as winter migrants 'had also been reduced'. The removal of the trees was of some long-term benefit to the farmers, therefore, for the stock doves had fed heavily on barley and turnips as well as beech mast. But White can add that in an atmosphere such as his at Selborne – where the air is generally moist and there are frequent 'night-time mists' – deciduous trees help to keep a locality watered. Smaller plants may be affected by trees in various respects, indeed, and may be largely dependent on them.

Some plants live only under trees or other cover, he points out. The stinking hellebore, *Helleborus foetidus*, thrives only in woodland; it flowers 'about January', when the trees are not in leaf. The bird's nest orchid, *Ophrys* (now *Neottia*) *nidus avis*, flowers in June or July, but it too was found 'in the *Long Lith* under the shady beeches among the dead leaves; in *Great Dorton* among the bushes, and on the *Hanger* plentifully' (under the beech trees).[19] The similarly named – but entirely distinct – yellow bird's nest, *Monotropa hypopitys*, lived in the shade of beech trees at Selborne, and 'seems to be parasitical' on their surface roots.[20] This is not correct, we know now, but the principle is correct: the yellow bird's nest is a saprophyte, and lives on and as it were through the leaf humus covering the tree roots. White seems not to have recognised the true parasitism of the toothwort, which takes its nourishment from the hazel. But one of the places in which he carefully locates the toothwort, we notice, is the Church-litten coppice: there it can be found 'under some hazels' and 'among the hazel stems'.[21]

Trees may be detrimental to the growth of other plants. When a section of beeches in the Hanger was cut down, wild strawberry plants sprang up in the newly exposed area, and four different sorts of thistles established themselves with the strawberries. Some of the vernacular names White uses for these thistles would not have been familiar to other botanists, but they would have understood him readily from the Linnaean binomials he also provides: *Carduus lanceolatus, nutans, crispus* and *palustris*.[22] The seeds of both strawberries and thistles had perhaps lain under the beeches for many years, he says, 'but could not vegetate till the sun & air were admitted'.[23] The seeds of the strawberries had been there, waiting their time, he had reason to think, since the same spot was previously cleared of trees.

The thistle seeds could, in fact, have been blown into the wood, and some of the strawberry seeds could have been dropped by birds. Though he does not refer to these modes of seed distribution here, he allows for them both in other contexts: for instance, he deduces that some beans appearing unexpectedly in one of his fields had been 'planted' there by jays. These beans were widely scattered, and

> It is most probable . . . that they were brought by birds, & in particular, by jays, & pies, who seem to have hid them among the grass, & moss, and then to have forgotten where they had stowed them. Some peas are also growing in the same situation, & probably under the same circumstances.[24]

William Derham had recorded that mice hide acorns one by one in pastures in the autumn, and White had watched rooks 'carrying off' acorns and walnuts from his own trees. (In a good year for fruit, small birds would descend on his soft fruit bushes 'in troops'.) A 'water rat' might gather potatoes into an underground store in a field, he could add, for once, while ploughing, a neighbour of his had once found 'above a gallon' of potatoes hidden in what seemed to be a rat's hibernaculum.[25]

The effects of weather are given the importance we would expect, in White's botanical notes. Often, events at Selborne did observe a settled regularity: his forecast for wheat and barley crops in 1774 is noteworthy for the local experience with which he backs it up. He writes this early in the year:

> We have had a very wet autumn and winter, so as to raise the springs to a pitch beyond anything since 1764 . . . The country people say that when the *lavants* rise corn will always be dear; meaning that when the earth is so glutted with water as to send forth springs on the downs and uplands . . . the corn-vales must be drowned: and so it has proved for these ten or eleven years past. For land-springs have never obtained more since the memory of man than during that period, nor has there been known a greater scarcity of all sorts of grain considering the great improvements of modern husbandry. Such a run of wet seasons a century ago would, I am persuaded, have occasioned a famine.[26]

But variations in the seasonal events of the parish – the weather, the 'first and last appearance' dates, the densities of given insects or the qualities of given crops – interested him as much as recurrences. Rain – or frost – early in the year may well affect a plant's eventual fruit- and seed-production, but by the summer and autumn the same plant will have been touched, to a greater or lesser degree, by many influences in addition to these. In the search for causal explanations, and as a more recent botanist puts it, 'opposites must be found, before discoveries can be made'.[27] The 1774 grain crops did turn out to be poor; but according to White's notes, the summer of 1775 was warm and dry by local standards, with periods of 'dripping', and though the previous winter and spring had been wet, the 1775 wheat crop at Selborne was a relatively good one. Frequently, by the use of 'opposites', he avoided false or premature 'discoveries'. At the beginning of September 1782, after a wet summer and 'contrary to all rule', the wheat in the parish was heavy and the straw short, whereas in 1781, after a hot, dry summer – and contrary if not to 'all rule' then to the expectations of several of White's friends – the wheat had been light and the straw long.[28] In 1784, after a wet summer, and following upon the exceptionally hot and arid summer of the previous year, 'No acorns, & very few beech-mast. No beech-mast last year, but acorns innumerable.'[29] In August 1783, he writes concerning the potato crop at Selborne: 'Potatoes very fine, tho' the ground has scarce ever been moistened since they were planted. They were also very good last year, tho' the summer was mostly wet & cold'.[30] Rain, it seems, is not of the first importance for the successful growing of potatoes. Gardeners may find this queer information; but perhaps, during the 1783 growing season, the 'dripping' phenomenon had again been playing its part.

A summer migrant does not arrive, a crop does not attain to a given weight, without a range of more or less influential concomitants; again to quote W. B. Turrill, a plant without an environment would be 'something not botanically recognisable'. Few eighteenth century botanists would have appreciated the point, even allowing for the awareness of 'economies', but White attended to environmental conditions respectfully if only because of his 'naturalist's apprenticeship'. He knew that trees and shrubs from other northern climates require careful locating in English gardens. Placed where they can receive the warmth of the sun directly, they will 'shoot away' at the first advances of spring, such is their native tendency, and may be cut off, therefore, by the frosty nights which can occur in Britain during March or April.[31] He cultivated sea-kale, not then thought of as a garden crop, from plants he brought back from Devon. Less successfully, he grew an *Arundo donax*, a giant reed, from seed sent him from Gibraltar. This reed reached a height of 'eight or nine feet', but when White opened a 'head' of the plant, in September 1775, after what had been a warm summer season for Hampshire, he found 'a long series of leaves enfolded one within the other to a most minute degree, but not the least rudiments of fructification'. Apparently, many more weeks of hot weather would have been needed to bring any fruits or seeds to maturity.[32]

But by a determined use of a purpose-built 'fruit wall', in particular, White ripened peaches, apricots and nectarines outdoors. In the shelter of a north wall, a showy 'Italian' plant 'called by my Grandmother Dragons, & by Linnaeus *Arum dracunculus*' had survived in the vicarage garden at Selborne 'thro' all the severe frosts of 80 or 90 years'.[33] He could himself *nurse* plants originating in countries to the south of Britain through even the colder Selborne winters. Usually, straw or twiggy branches placed over grape vines, arbutuses or Portugal laurels would save their new shoots from the effects of frost, and during a sudden cold spell in March 1766, we read, 'the wall-trees have been boarded, & matted all day'.[34] In the winter snow he regularly experienced, and with his vines and some of his evergreens, shaking the snow from their twigs or foliage after each fresh snowfall helped to preserve the tender buds. A repeated thawing and re-freezing, he says, for instance as day and night temperatures follow each other, is especially detrimental to plant life.[35]

He became a hot-bed expert, we saw. Long before he had seen his natural history role plainly, he was testing a garden soil to its limits; for the botanist he became, the effects on plants of *soils and terrains* were matters for continual enquiry. Living in a 'relatively high' region, he had no immediate access to large rivers, and was without 'large aquatic plants', but this defect apart, the great range of soils in the parish, and its contrasting – exposed and sheltered – situations, guaranteed the botanist an 'ample flora'. The letter (DB xli) *following* the letter in which he touches on classification shows us White the botanist at work. (Of the twenty plants referred to in DB xli, only one of them, the stinking hellebore, is singled out as of use to man – and it is of dubious use.)[36] The cranberry or creeping bilberry, *Vaccinium oxycoccos*, grows in the boggy stretches of Bins Pond, we are told here, and the related whortleberry, *Vaccinium myrtillus*, thrives on the 'dry hillocks' of Woolmer Forest. The opposite-leaved golden saxifrage, *Chrysosplenium oppositifolium*, can be found in the 'dark and rocky hollow lanes' worn into parts of the parish landscape, and the similarly 'rare' autumn lady's tresses – White gives the original Linnaean binomial, *Ophrys spiralis* – grows on the grass-topped Long Lythe and on one corner of Selborne Common.

The rubbly 'white land' of the parish was fit for neither pasture nor the plough, but was 'kindly' for hops. The upland chalk and the freestone were particularly suitable for beech trees, so that the local beech woods were, as they still are, often 'hanging woods'.[37] Soils make the great difference between Woolmer and the Holt, the two adjacent 'forest' tracts, White says; the soil of the Holt, which carries a good turf and abounds with fine oaks, is 'a strong loam, of a miry nature', while that of the 'treeless' Woolmer is almost entirely a sand.[38] He does not jump to conclusions regarding plants and soils. A large oak grew in front of the terrace at The Wakes, after all, on more or less chalky ground. Finding a plant in particular conditions, he asks which of these conditions seem to facilitate its growth and maturing, or where appropriate, which restrict this development. Timber was of constant importance in his time and place, and he refers to 'shaky' timber from time to time. The effect of shakiness, and the first indication one has that a tree is shaky, is that when the timber is sawn it cracks along lines concentric with the heart, making it unsuitable for most kinds of joinery. He assumes initially that this undesirable quality relates to the soil in which the tree has grown: in the Selborne parish oaks were numerous on both the clay and the freestone rubble, but on the latter, though they sometimes reached a large size, they were often marred as timber by this shaky brittleness. But at least with chestnut trees showing the same defect, he has to qualify this view. Shaky chestnut timber had frequently been taken from Rotherfield Park, on the other side of Newton Valence, where the soil is predominantly a chalk. (John Carpenter did have a use for some of this timber, it made light buckets and tubs.) But a stand of chestnuts felled at Bramshott, beyond Woolmer Forest, showed the same readiness to crack and flake, and the soil at Bramshott was a 'sandy loam'. In chestnuts, he decides therefore, this troublesome defect may relate as much to the sort of tree as to the soil the tree has grown in.[39]

A regular exchanging of seeds and seedlings was a feature of White's gardening life, where this often involved a move from one soil to another. He could compare his garden plants with garden plants at the village of Bramshott, which lay seven miles to the east of Selborne, in the valley of a tributary of the Wey. From the late 1770s onwards he visited Bramshott Place, the home of a Mr Richardson, at some time during the summer or autumn of most years. With the aid of liming and turniping, he found, the Bramshott soil – the sandy loam – would produce as good a corn crop as any at Selborne. In July 1781, 'Mr R's garden, tho' a sand, abounds in fruit, & in all manner of good & forward kitchen-crops'. The same soil did not 'clod' and crack intolerably in periods of drought, and after a visit to Mr Richardson's during the summer of 1783, White asserts frankly, 'Sandy soil much better for garden-crops than chalky'.[40] Average temperatures at Bramshott were higher than those at Selborne; the garden produce there might be ahead of White's own by as much as a fortnight in spring, and the fruit ripened earlier than his. Because of its special advantages, indeed,

Bramshott Place could usually show 'an abundance of everything'. Usually, but not quite invariably. In June 1791 (towards the end of his life), the visiting student could record; 'Mr Richardson's straw-berries very dry, & tasteless'.[41]

But how one observes is as important as where one observes, White knew. He could challenge the merely physiological botanists and the 'mere classifiers' without leaving his own garden. In the letter to Barrington on plants and the conditions in which they flourish (DB xli), the most striking item appears towards the end. The 'spring' and 'autumn' crocuses, common though they may be, are especially clear examples of 'congenerous species with divergent modes of life'. Even the best botanists can discover no differences between the two 'in the corolla, or in the internal structure', he writes, and yet the vernal crocus is in flower by the beginning of March at the latest, frequently in 'very rigorous weather'. By contrast, the autumn crocus, ignoring the influence of spring and summer, does not flower until many other plants are fading or running to seed. Common events, he adds, can be as thought-provoking as the most stupendous phenomenon in nature.[42]

3 Minute but multitudinous beings

The crickets White studied gave him examples of diverse manners and habitats in creatures of the same genus, we said.[43] Insects – he uses the term broadly – do not occupy a major place in his enquiries, and as it happens, insect life had its dedicated observational students even in the eighteenth century. Colonial insects, in particular, can in many cases easily be kept and bred in conditions which allow close observation. In White's time the most outstanding contributor to behavioural entomology – and the most outstanding user of 'dioptric' bee-hives – was Reaumur, although partial parallels appeared in England in the work of Wildman on bees and William Gough on ants.[44] Observational and environmental, White was aware of the total parish, as well as of particular events within it. He nowhere exemplifies the exclusive specialisation typical of the great insect watchers, of his own or the succeeding century. But where he could, he lived on equable terms with insects. Where insect-watching offered itself, he willingly accepted the offer; 'I must not pretend to great skill in entomology', he tells Thomas Pennant,[45] but insects are the subjects of some of his most impressive brief portrayals.

The typical entomologist of the time was, again, a more or less knowledgeable collector. By even the 1750s the enthusiast's equipment might include,

> at least two kinds of nets, beating-sticks and -sheets, pill-boxes (for putting over insects when captured), cork-lined pocket collecting-boxes, pincushions (for holding the pins, in the days when specimens were always 'pinned' in the field), cork setting boards, card 'braces' for setting, setting-needles, breeding cages and even special trowels,

for digging around the roots of trees for pupae.[46] While ornithological and botanical collectors are at least noticed in the papers White prepared for publication, the entomological collector, including the lepidopterist, is not once referred to. Again, this silence did not represent mere innocence on White's part, although he was sometimes satisfied to convey this idea. His reply to Barrington, when the latter asked him to procure a live blackcap warbler, was, 'I am so little used to birds in a cage, that I fear if I had one it would soon die for want of skill in feeding'.[47] In truth, if anyone could have kept a blackcap alive in a cage – a difficult feat enough – it was Gilbert White. He was familiar, he had familiarised himself, with the ways of collecting and preserving insects. In one of the letters to John White, he passes on details:

> Many water-insects to be taken with a spoon-net. Mr Drury will send you a hand-fly-trap on the principle of a pair of snuffers: the bows are lined with gauze; and a clap-net. Take libelellulae, dragon flies: cut off the abdomen from the thorax, & taking out the bowels draw the empty body on upon a straw that is fit; or fill the cavity with chalk.[48]

He had been determined to help the luckless John, and in the Gibraltar correspondence he was himself still getting his bearings; but by the time the correspondence with Pennant and Barrington was fairly begun he was committed to behavioural investigation. In *Selborne* and the *Journal* – instead of accounts of 'spoon-nets' and pill boxes – he gives us delicately focused close-ups of insect activity.

The *Bombylius medius*, a hovering, darting bee-fly, was often watched during March: 'The female seems to lay it's eggs as it poises on it's wings, by striking it's tail on the ground, & against the grass that stands in it's way in a quick manner for several times together'.[49] The drama of the insect world was not lost on this intimate observer. In September 1784,

> I lately saw a small Icneumon-fly attack a spider much larger than itself on a grass-walk. When the spider made any resistance, the

> Ich: applied her tail to him, & stung him with great vehemence, so that he soon became dead & motionless. The Ich: then running backward drew her prey nimbly over the walk into the standing grass. The spider would be deposited in some hole where the Ich: would lay some eggs; & as soon as the eggs hatch'd, the carcase would afford ready food for the maggots. Perhaps some of the eggs might be injected into the body of the spider, in the act of stinging.[50]

Ichneumons will sometimes deposit their eggs in the pupa of a butterfly or moth, he adds: at an earlier date he had viewed this exploitative behaviour, too, at first-hand.

As well as the bee-like *Bombylius*, he describes various true bees at work. *Eucera longicornis*, a black, hairy burrowing bee, bored into the hard soil of his garden paths in June; in July it 'carries wax on it's thighs into it's hole in the walks', and then deposits it's eggs there.[51] A carder bee, *Anthidium manicatum*, is caught in the act of neatly shaving the fibrous outer covering from a twig. 'When it has got a vast bundle, almost as large as itself, it flies away, holding it secure between it's chin and it's fore legs'; it will use this material for nesting purposes.[52] In June 1776, the especial behaviour of a 'vast' moth was recorded. It 'appears after it is dusk, flying with a humming noise, & inserting it's tongue into the bloom of the honey-suckle': it scarcely settles on the plants, but 'feeds on the wing in the manner of humming birds'.[53]

His most notable piece of detective work with insects was that relating to a black, 'dust-like material' on one of his vines, which he investigated in 1785. He applied a glass to this dust, and could find nothing in it that seemed animate; but on looking behind the boughs of the vine he found them 'coated over with husky shells, from whose sides proceeded a cotton-like substance, surrounding a multitude of eggs'. He recalled having heard or read of something similar, and with the help of his books he identified what he was looking at as the European vine coccus, although he was unaware of this dangerous pest's having ever previously appeared in England. The coccus must have been sent to Selborne alive and inadvertently, in one of John White's parcels of plant and bird specimens, he decides; insects may travel from one country to another in unexpected ways, and in some cases are 'wonderfully hardy'. He had noticed the 'dust' during a number of previous years, so that wherever the original insects came from, they or their descendants had already survived cold winters in their present situation.[54] According to Walter Johnson, who looks at the episode in detail, White's history of this coccus was an advance on that of Reaumur and can be called the first accurate history.[55] Without doubt, it was the first accurate report of the insect in the British Isles.

That one function of colour and scent in flowering plants is to attract insects (or that, in a sense, plants send signals), is not emphasised by White, and communication between insects is not considered at length by him. But he did try to find out whether *bees* could *hear* – this although the experts he was aware of had ruled out hearing in insects, insects having 'no auditory organs'.[56] As with the young chicks to which he presented wasps and flies, and the young vipers taken from the body of their mother, his elementary work on insects and hearing foreshadows uncannily work which was to be performed many years later. He shouted at the bees through a speaking trumpet while standing close to the hive, to begin with, and they seemed not to react. His first conclusion, therefore, was that they could not hear.[57] (He might have been warned of the limitations of a test of this sort by his own earlier experience: he had tried the same thing with the tortoise, and had noted on that occasion, 'When we call loudly thro' the speaking-trumpet to Timothy, he does not seem to regard the noise'.)[58] The creatures had no reason to react, we can suggest. When Keller tested fishes in an aquarium, in 1883, by getting someone to sing loudly beside them, they 'seemed not to hear', but when on subsequent occasions he fed the fishes as the singing began, and thus gave the disturbance a relevance for them, they soon swam to the surface of the tank merely on the commencement of the singing. White, however, could not believe that the chirruping or squeaking of insects was produced quite gratuitously, or merely because of its effect on other orders of creatures, and he knew that insects sometimes responded to events they could not see. The 'singing' of male field crickets served much the same purpose as the singing of cock birds, he decided; it had to do with, for one thing, 'rivalry and emulation'. The more sociable house crickets, again, if they are surprised by a candle while out in a room at night, 'give two or three shrill notes, as it were a signal to their fellows, that they may escape to their crannies & lurking holes to avoid danger'.[59] Insects are delicately sensitive to vibration, he comes to think. If they do not hear, they may be able to '*feel* the percussion' of sounds.[60] The nineteenth century discovery that insects have highly acute organs of auditory reception would not have surprised him,[61] and the more recent discovery, that various cricket species have auditory receptors on their 'elbows' and 'knees', we can be sure, would have greatly encouraged him.

He largely passes over butterflies; they were already favourites of the collectors; though he does not fail to record the occasional swallowtail; one presented itself at The Wakes in August 1780, and he adds, 'In Essex & Sussex they are more common'.[62] He seems not to have kept caterpillars or other larvae, to watch their transformations, but he describes the way he induced field crickets to come out of their burrows so that he could examine them in the hand. Attempts to dig these insects out merely crushed them, he says, but a 'pliant stalk of grass' twisted into the burrows brought them out alive and in good order. (Using much the same method, Jean Henri Fabre would get even *Lycosa* spiders out of their holes). By this means, White says, 'the humane enquirer may gratify his curiosity without injuring the object of it'.[63]

White could be temporarily seduced into non-comparative observing, by insects. He writes sometimes as if the neat and cleanly field crickets, or the sly house crickets singing and feasting by a Christmas fire, belonged to small, independent 'economies'.[64] A counter-tendency, and a quite unsentimental approach, was encouraged by his being a dedicated gardener and small farmer. In these capacities, he distinguished between useful, harmless and 'noxious' insects – and was sure the last group should be vigorously combatted. The vine coccus, potentially a 'loathsome' pest, was an unexpected arrival at Selborne, but in a country parish in White's time devastating insect damage could occur during almost any summer. Turnip and cabbage 'flies' might infest the field crops in such numbers that they 'pattered like rain' when disturbed, and depending on the year, kitchen gardens might be ravaged by summer chafers.[65] Periodically, the caterpillars of a small 'yellow' moth swarmed on the hedges and trees in mid-summer, destroying the foliage of 'whole forests and districts': in June 1785, 'most of the oaks' in the Selborne parish were quite leafless from this cause.[66] A complete history of insects hurtful in the field, the garden and the house would be 'a most useful and important work', he declares in *Selborne*, therefore; 'a knowledge of the properties, oeconomy, propagation, and in short the life and conversation of these animals, is a necessary step to lead us to some method of preventing their depredations.'[67] He himself directed annual campaigns against wasps. His forces were again drawn from among the younger members of the village community, but he records towards the end of the 1785 season, 'Boys bring the 26th wasp's nest'.

He knew the turnip fly to be 'a leaf-beetle of the *Chrysomelae family'*, and he was clear that the maggots which spoiled gammons and hams as they hung in the village chimneys belonged to 'a variety of the *musca putris* of *Linnaeus*'; but with insects more than any other creatures, the investigation of their lives was often obstructed by identification problems. In 1770 there was still 'no good English Hist: of Insects', or no history that began to be comprehensive. English names had still to be found for 'half' the insects of Great Britain, and just as importantly, 'neat plates' which would clearly indicate 'the *genetic distinctions* of insects according to *Linnaeus*' were generally wanting.[68] White thought the large, diagrammatic engravings in the *Histoire abrégée des insectes* (1762), by E.L. Geoffroy, were some of the best of their kind. This work was not translated into English, but he sent a copy of it to John White. He refers to it himself in connection with the 'star-tailed', water-dwelling larva of *Stratiomys chamaeleon*, or as he calls it also, the chameleon fly.[69]

Noxious or benign, and whether or not they appreciate the larger world they are part of, insects are adapted to their surroundings. The mole cricket's 'fore-feet' are curiously suited to underground digging (they are rather like the fore-feet of the mole), and in field crickets, the jaws with which they excavate their regular burrows or seize an insect enemy are 'toothed like the shears of a lobster's claws'.[70] This equipment had been given these creatures, or species, White assumed; he seems not to have associated it with a process of gradual adjustment. But he generally concerned himself with his subjects' habits and social affiliations; he knew that what insects did habitually was adapted, and, above all as someone living in a country house, knew that the doings of insects too might 'vary and conform to place and convenience'. The house crickets continued to reproduce in winter, when other comparable animals were either dead or in a state of diapause, we saw. The same successful insects were more or less omnivorous; they would eat salt, yeast and any kind of kitchen offal and were lovers of milk, broth and beer; they gnawed holes in woollen stockings hung near the fire to dry, and sucked the pulp from some oranges.[71] A colony of 'black ants' lived under the floor beneath the staircase at The Wakes. In July or August, when the large winged females and smaller males emerged for their departure, they temporarily 'filled the rooms'. On these occasions, agitated worker ants also showed themselves in great numbers, and workers might appear in twos and threes even in December. How can the colony continue, the naturalist asks, with 'no communication' with the garden or yard?[72]

Cockroaches came to his notice when they appeared at Selborne for the first time in 1790, and soon established themselves at The Wakes. Here a campaign was mounted by Mrs John White, who was a widow by this time and had been given a home by Gilbert; but as fast as she destroyed these creatures their numbers were made up by recruits from other houses. 'These, like many insects,' White observes, 'when they find their present abodes over-stocked, have powers of migrating to fresh quarters'.[73]

He considered and almost identified the maggots which left worm holes in chairs and bed posts, and various more obviously parasitical insects appear frequently in his writings. Horse bots and 'harvest mites' were persistent country problems,[74] and to these, and the warble and ichneumon flies, could be added the lice and dipterous insects which infested birds, and most notably in his experience, the hirundines. The historical, evolutionary dependence of plants on insects was not suspected by him, but the symbiosis of plants and insects in the present state of things was manifest, at least to a Gilbert White. 'The day and night insects occupy the annuals alternately', he notes during one September, 'the papilios, muscae. & apes are succeeded at the close of the day by phalenae, earwigs, woodlice, &c.'[75] Insects are a massive source of food for other animals, and as food are a governing factor in the lives of many migratory birds, and insects in their turn are controlled by birds in particular. White watched an instance of this controlling one day in May, perhaps from his garden. 'The crows, rooks, & daws in great numbers continue to devour the chafers on the hanger', he writes; 'Was it not for those birds, chafers would destroy everything.'[76]

His notion, whether original or adopted, of how flies stick to smooth vertical surfaces, was mistaken.[77] He returns at several points to the clammy, sweet-smelling substance which disfigured various of his flowering shrubs in hot weather, the so-called 'honey-dew', and here too he accepts a far-fetched explanation: 'in hot days the effluvia of flowers are drawn up by a brisk evaporation; and then at night fall down with the dews, with which they are entangled'.[78] The honey-dew material is not 'of the vegetable kind', it is produced by aphids, we now understand. But – as with the house flies, which White tells us have 'flat feet' – he had gathered some of the pertinent data. In the *Selborne* references to honey-dews, he especially notes the presence of aphids. In 1783, in May, he remarks in the *Journal*: 'Sprinkled & washed the foliage of the fruit-trees, that were honey-dewed, & began to be affected with aphides'.[79] Tar water, incidentally, seems to have been his chief insecticide.

He knew nothing of the wonderfully involved life cycles of many of the aphid species, which are today still being uncovered, but he records enormous fluctuations in insect populations. When during one summer no wasps were to be seen, he found it extraordinary that the species could survive: he was aware of the 'big breeders', the queens, but he may not have known that only one of these queens is needed to found a new nest colony. He notes that while wasps are attracted to our orchards, kitchens and butchers' shops when their larvae are hatched, they nest also in 'woody, wild districts far from neighbourhoods'; there they feed on the nectar of flowers, and catch flies and caterpillars to take to their young.[80] (Wasps are vegetarian, he knows, but their larvae are not.)

The turnip and cabbage flies, if they pattered like rain 'during hot summers', did not do so during all hot summers; but insects were often delicately responsive to weather conditions. He writes in January 1771:

> Wood-lice, onisci aselli, appear all the winter in mild weather: spiders appear all the winter in moist weather; lepismae appear all the winter round hearths & in warm places. Some kinds of gnats appear all the winter in mild weather, as do earth-worms, after it is dark, when there is no frost.[81]

The recording of relations is unending in Gilbert White; if he had had a personal computer, it would have been in daily use.

Seasonal movements of insects are touched on, particularly in the journal/notebooks. Field crickets were vulnerable when taken 'out of their knowledge', but house crickets, like cockroaches, flew out of the windows on warm summer evenings and might then take up residence in houses they had not previously used. The true complexity of insect colonies was barely approached in White's time (despite the dioptric bee hives), but the sudden mass emergence of flying ants in July or August, he knew, while it encouraged the hirundines among other birds to 'feed deliciously', allowed the female ants, or those which escaped, to scatter widely before founding new ant cities. With the help of strong winds, aphids sometimes emigrated in huge numbers. An 'aphides shower' fell on the Selborne parish in August 1785; these insects had come, he suggests, from the great hop-fields of Kent or Sussex.[82] Memorably, on a clear, calm day in September 1741, myriads of tiny spiders had sailed over White's part of Hampshire. These spiders were carried by 'rags of cobweb' which twinkled like stars while aloft and coated the meadows and hedges when they fell to earth. The episode was unique in the naturalist's experience, and remained unexplained; although he was aware that on any fine day in autumn, in his parish, small spiders were 'shooting out their webs' and taking to the air to be locally transported by this means.[83]

Chapter 8

THE USEFUL NATURALIST

1 'Improvement' at Selborne

A special respect was reserved by White for the hard working and provident creatures among his parish fauna. The unwearying swallow, collecting food for her young ones throughout the hours of daylight, is an 'instructive pattern' of loyalty and industry, he says. In a different but not altogether unrelated mood, and in the interest of his gardening and that of his neighbours, he hunted out wasps and, it has to be added, shot the birds which plundered his fruit bushes; or had them shot, for he generally passed on this task to one or other of his helpers.[1]

He dwells periodically on the 'folk wisdom' of his part of Hampshire, and leaves notes on various of the received agricultural practices of his region and time; in our assessment of these practices, he says, utility must be the main consideration. The management of pastureland, for instance, was often ill-informed and wasteful. The graziers and farmers 'do not seem to distinguish any one sort of Gramen from the other – the annual from the perennial, the succulent from the dry, or the aquatic from the upland; whereas by attention their meadows and pastures might be much improved'.[2]

He made annual journeys to Ringmer, near Lewes, for many years, along the South Downs, and became aware of a tradition accepted by the sheep farmers who used the Downs. To the east of the River Adur they kept a breed of hornless sheep, with black faces, a tuft of wool on the forehead and black-spotted legs; and to the west of this river, and back as far as Chichester, they reared only horned sheep, with white faces and white legs. The 'east country poll-sheep' produced the finer wool, White discovered; and in the manner we come to expect of him, he asked the shepherds why this strict separation was adhered to. They tell you, he writes, that 'the case has been so from time immemorial; and smile at your simplicity if you ask them whether the situation of these two different breeds might not be reversed'.[3] His friend in the Chichester district (John Woods), he goes on, however, has decided to run some black-faced hornless rams with his horned western ewes in the autumn. White had first noted the shepherds' preference or prejudice some years earlier, in 1769, and it is likely that he suggested the 'experiment' to his friend. At all events, in August 1776 he records in the *Journal*, 'Mr Woods of Chilgrove thinks he improves his flock by turning the east-country poll-rams among his horned ewes.'[4]

The 'agricultural revolution' got fully into its stride, during White's adult lifetime. At Selborne the farming was 'mixed', and was in a state of transition; moderate sized farms and mere copy-holdings, traditional methods and new and progressive techniques, were observable side by side. Seed 'drilling', the planting of field crops by horse-drawn machine, was in use on some of the farms, although as White notes, this procedure could be difficult on dry or binding ground.[5] Reducing the surface soil to an open tilth was vital to successful cultivation (this was a problem he had personal experience of). He comments in November 1781:

> Husbandry seems to be much improved at Selborne within these 20 years, & their crops of wheat are generally better: not that they plough oftener, or perhaps manure more than they did formerly; but from the more frequent harrowings & draggings now in use, which pulverize our strong soil, & render it more fertile than any other expedient yet in practice.[6]

White, though he farmed himself in only a small way, was familiar with the business aspects of farming; but his attitude was most obviously that of the working owner commended by William Marshall, in his *Minutes of Agriculture* (1787). 'He esteems self-attendance and close attention, even to the merest minutiae, absolutely necessary to common management; nor thinks the manual operation of the humblest department beneath the man.'[7] White records the new procedures of clamping turnips in autumn and of stacking rather than 'housing' wheat prior to its being threshed; he refers to the danger of fire if hay is carried and stacked prematurely.[8] He speaks confidently of *under-sowing* a cereal crop; Farmer Knight, who found it hard work to plough Baker's Hill in the entry we looked at earlier, was to sow the same ground with oats, '& I am to sow a Crop of St.foin along with his Corn'.[9] This was another of the involved co-operative arrangements typical of the younger White, and no doubt his Selborne neighbours, too, sometimes 'smiled at his simplicity'. (To grow sainfoin at all was still something of an innovation.) But he encouraged the adoption and growing of potatoes in the Selborne locality with great success.[10] He made practical suggestions concerning the cultivation of hops, and grew some 'American wheat' in his Pound field of a sort which was not easily beaten down by wind or rain. Wheat and barley grew much taller, the straw was much longer, in his day than is now the case, and 'Wheat is so apt to lodge in these parts, that they are often obliged to mow it down in the blade about May, lest it should fall flat to the ground.'[11] The present-day corn grower, even with his shorter straw, would find this a familiar topic.

White knew the importance of appropriate manures and soil conditioners, we have seen. 'Turniping', the growing of turnips on poor ground to be eaten in the field by sheep which in their turn would dung the ground, was in use in the parish, but though the farmers varied their manuring and soil treatments (their liming, for example) they varied them, it is evident, in a hit and miss fashion. New methods of cultivation and new crops may be required, but they should be tried and introduced methodically, White implies. Note-taking will be indispensable; he illustrates the approach in the *Journal*, his own general fieldbook. (Here, he does in occasional notes what William Marshall would do comprehensively in his *Minutes*.) In February 1771, 'Bro^{r} Henry's field opposite his house, was fallowed for barley before the two frosts, all save the headlands: mem: to enquire if the earlier fallowing in that part proved to any advantage.'[12] Early in March of the same year, 'Farmer Parsons sows wheat in his fallow behind Beacher's shop, which was drowned in the winter. Mem: to observe what crop he gets from this spring-sowing.'[13] Or in a retrospective note added to the *Journal* entry for April 1st 1771; 'Mr Woods, of Chilgrove, had on this day 27 acres of spring-sown wheat not then sprouted out of the ground: & yet he had a good crop from those fields, no less than 4 quarters on the acre!'[14]

Sainfoin, a vetch, was one of various 'new' multi-purpose crops; it both provided winter fodder and improved the ground. It was grown by White for many years – it was usually bought by Timothy Turner, the smallholder on various pieces of village land – as an experiment and an example.[15] In 1792, the year before he died, White and his brother Thomas were applying different treatments to each of four comparable sections of one of his fields, using lime, gypsum, wood and peat ashes and lastly peat dust, all in measured quantities.[16] Gilbert does not refer to Jethro Tull (1674–1741), the pioneer of continuous cultivation, or his contemporary 'Turnip' Townshend. He says nothing directly about Bakewell (1725–95), the scientific stock-breeder, or Coke the transformer of poor land at Holkham.[17] But he was on the side of these reformers, albeit with some reservations. Even the Selborne parish, if it was to be what it ought to be, needed the 'forming' hand of the enlightened agriculturist and planner.[18] White's extensive use of garden glass reflected contemporary trends, we said. He refers to 'noble' greenhouses, and he leaves an awed description of a greenhouse being prefabricated at South Lambeth, to be shipped to Naples:

> The whole area of this house will contain 4680 square feet, & the two beds, or borders 2200 square feet. As the soft & southern climate of Naples produces lemons, pomgranades, citrons, & many trees & plants with which we croud our green-houses; we are to suppose that this royal Conservatory will be furnished only with the most fragrant, choice, & rare vegetable productions of the Tropics.[19]

He nowhere mentions Alexander Kent or 'Capability' Brown (1715–83), but he was 'an admirer of prospects'. (He regretted that the Selborne church was without an elegant spire.)[20] On the evidence of his notebooks and what can still be seen – he not only used hot-beds and frames successfully but built a ha-ha and laid out 'walks' – it is plain that he too found 'capability of improvement' in his estate.[21]

White was not aping the aristocratic landowners and great gardeners, or at least, was not trying to pass himself off as grander than he was. (He *tells us* that the figure of Hercules had been made for him by John Carpenter the village carpenter.) He believed that even the humblest could play their parts, rather; Selborne need not be ashamed of itself; and like the leaders of the improving 'movement', he was an optimist. He began methodical natural history work of his own – to distinguish it from the rather different work he did with John White – when he was in his mid-forties, and he started building the 'great parlour' when he was fifty six, an advanced age in the eighteenth century. His most startling horticultural feats were performed during the 'early' gardening years, but in 1787, when he was sixty six, he cut two frame cucumbers on April 18th.

The good man, he was sure, was a 'good commonwealth's man'. The gardening suggestions included in *Selborne* would be of practical use, and in his paper on the making of rush lights he was both recording details as an observer and providing information a benevolent landlord might pass on to 'the poor'. He and his brothers planted trees about the village, although the four limes planted by Gilbert in the village street, opposite The Wakes, to 'screen off the butcher's yard', were no doubt for his own benefit first of all. Again with family help, he tried sowing the south-eastern side of Selborne Down with quantities of beech mast and holly berries, and constructed permanent paths on the Down.[22] He took an active interest in the 'roads' of the parish, which were dusty in summer and so muddy in winter that a chaise could not be brought into the village proper. Towards the end of the *Garden Calendar* he notes that he has had the stream below Gracious Street cleared of the accumulated mud which caused the lane there to be flooded during any wet season.[23] At about the same time, he had some seventy yards of cobbled pavement laid along part of the 'straggling' village street.[24]

2 The historical dimension

The Selborne relationship, White's sense of Selborne, was such that what threatened the parish caused him personal anxiety and what redounded to its credit was for him a source of pride, but other Selbornes, he knew, had preceded the eighteenth century parish.[25] He tried to tease out the archaic origins of local place names. The history of the Hampshire dialect could be studied, he says, if one 'frequented farmers' kitchens'.[26]

In former times, he notes in one of the *Selborne* papers, quantities of ancient, blackened timber, so-called 'fossil wood', were taken from the bogs of Woolmer Forest; in a later paper he notes that the peat-cutters still occasionally find pieces of this preserved timber. Woolmer had been virtually without trees for some centuries, he knew, but trees had once grown there. From his examination of some of the fossil timber, he gathered that a proportion of the trees had been deciduous; a recently recovered piece might have been the butt end of a small oak – and it had 'apparently been severed from the ground by an axe'.[27] White did not believe there had been a Roman settlement on the Forest; but at some earlier time, clearly, this large sandy tract had been more productive than it now was.

His own part of the parish, the 'abrupt, uneven' part in which the village proper was situated, was relatively well wooded, but he knew that many more trees had grown there in earlier times than could be seen in the 1770s and '80s. One strong indication of this, he thought, was that there had once been at least three water mills along the Selborne stream; by some means, the stream had once been more plentifully supplied with water than it was in his day.[28]

Three of these mills had been controlled originally by the Selborne priory.[29] White was vague regarding the pre-Christian history of his region, but with the help of Dr Richard Chandler, an antiquarian and friend, and over a period of some ten years, he made a close study of the documents held at Magdalen College, Oxford, on medieval Selborne. The lengthy appendix to the *Natural History* as first published, the *Antiquities*, was the result of this research. The earlier Selborne residents had without doubt been 'as busy and bustling, and as important', he says, as the parishioners of his own time.[30]

The Selborne manor-house, the priory ruins, the copy-holdings and the fuel and foraging rights were all links with this previous era. Antiquarian research was in vogue, and between 1777 and 1785 especially, the attention White gave his antiquarian studies reminds us almost of the attention his gardening received during the 1750s. His delving into the past of the parish held up the publication of *Selborne,* and for a time threatened the general character of the book. (For a time, he seems to have toyed with a notion of 'parochial history' which, had he adopted it, would have brought his work closer, at least, to that of Plot and William Borlase.[31]) But the *Journal* records continue uninterrupted throughout this period, and show little sign of the infatuation, and in the event *Selborne* remained a field naturalist's book. If White was mistaken in tacking the often discursive *Antiquities* onto the end of *Selborne* – and one object even in the *Antiquities* was the uncovering of 'customs and manners' – he was right to insist that a serious examination of the parish had to allow for its deep historical roots.[32]

'Priory Farm' and 'Grange Farm' recalled this earlier age, if only because of their names, and any of the farms of the parish might have recalled it – because of their modest size. We have the advantage here of being able to view the Selborne of White's time from a distance. Referring to the varied complex of land holdings which still obtained in the eighteenth century parish, W. S. Scott points out that there was no 'squire' of Selborne.[33] The priory had had its lands from the Saxon manor, and on the dissolution of the priory, in the thirteenth century, its property had passed to Magdalen College. Magdalen was still the landlord of considerable parts of the parish in the eighteenth century, with this important consequence: *new* large estates could *not* be established there.[34] In central and southern England, the processes of amalgamation and 'inclosure' were given a new impetus, during the second half of the eighteen century; they were encouraged by the great agricultural reformers themselves, and could affect holdings of all kinds. Towards the end of White's life, he helped defeat an attempt to inclose Selborne Common, on the top of Selborne Down; but at Selborne such an attempt was exceptional. The farms of the parish, if they were sold at all, went to separate owners, and the Common and the copy-holdings, the legal access of the village people to the 'common fields', remained intact.[35] In April 1784, when some oaks were felled in the Holt forest, and Lord Stawell, the grantee, tried to deny the cottagers the lop and top firewood they claimed as of right, there was a 'riot', White tells us; 'these folks, especially the females . . . set my lord at defiance'.[36]

But Selborne was altering, even though the agricultural pattern remained 'diversified'; and if White was an antiquarian in one of his hats, he was not nostalgic for an earlier order of things. The 'great improvements of modern husbandry' had kept up corn production even during the succession of wet summers (from 1764 to 1774),[37] and in the 'progressive' manner, he recognised that benefits were arising even from the inclosure process. Most remarkably, the quantity and improved quality of crops grown for winter fodder meant that fresh meat was available even during the winter months.[38] One has to be able to afford the fresh meat, White agrees, but he points confidently to 'modern' changes even in the lives of the poorest people. The wheaten bread replacing the inferior sort made from barley or beans was helping to reduce the skin diseases so prevalent in earlier ages, and so was the use of linen shifts and shirts in place of 'filthy woollen'. By his own time, 'every decent labourer also has his garden', and even common farmers provided their living-in workmen with greens and pulses to eat with their bacon. White's view of modern developments might have been less sanguine if he had lived twenty or thirty years longer; but three quarters of the way through the eighteenth century, these changes could seem promising indicators for the future.[39]

White's remarks on the Downland sheep apart, we hear little from him about the effects on farm stock of selective breeding. Breeding techniques which had long been used with dogs and race horses were applied to cattle, sheep and pigs with extraordinary success during his century. But here too he is cautious rather than dismissive. What he disassociates himself from is not the artificial but the 'whimsically' artificial.[40] The exhibiting of a cage bird in a glass chamber, itself set inside a goldfish bowl, epitomised this whimsicality, and he perhaps placed Barrington's experiments with song birds in the same category. An interference with nature could easily be *de*grading to the creatures concerned. A rider accompanies his rhetorical invitation to the landscape gardeners and farmers who would find scope for their talents in the Selborne district, therefore. They should by no means exemplify that 'unpleasing, tasteless, impotent' self-assertion which reduces a garden to 'flats, with loads of ornaments supply'd';[41] they were not to impose themselves arbitrarily on the landscape or its human and non-human inhabitants. A proper management on the part of gardener or farmer presupposed a detailed knowledge, in any instance, of that which was to be managed. One gathered from the study of a piece of land where and how it should be improved.[42]

White was an Anglican optimist. Human beings were the delegated agents of the Almighty; their business on earth was to help run and even complete a natural world which had been 'created' but which, for whatever reason, was imperfect or incomplete. He was glad to be able to tell his correspondent Robert Marsham, in 1789, that 'Jacobin clubs' were unknown in his part of the country, but he believed that God had bestowed 'a degree of reason on all his domestics . . . without distinction'.[43] He was without condescension in his relationships with his fellow villagers, but he it was who taught Thomas Hoar to read and write. If he approved of mutual support within the parish, he also encouraged those parish members (such as Timothy Turner) who were not only prudent but relatively ambitious. His own gardening and farming were in part a demonstration of what could be achieved when 'modern' horticultural aims were allied with 'natural knowledge'. For the thinking White, even a sociable utility was an insufficient standard on its own; the responsible landowner needed some knowledge of modern technology, and as well as this, a familiarity with advances in both botany and zoology.

White could repeatedly change hats. As a parishioner, he handled and in a sense interfered on principle, but as a naturalist and scientist he subscribed to impartiality as an ideal. Impartiality by no means entails passivity, but generally he made his natural history contribution without also physically interfering.

Chapter 9

SCIENCE, METEOROLOGY AND GEOLOGY

1 More quantities than qualities

White believed that outdoor nature study would be useful, but he did not verify as a naturalist by appeal to usefulness. His 'natural knowledge' was scientific in a sense the present-day biologist can recognise. It was valid in so far as it provided qualified predictions – in such and such circumstances, such and such further events can be confidently expected – and in so far as it survived the testing of these predictions in the field.

The same natural knowledge was 'non-metaphysical', and depended on public evidence; its claims were of a kind other naturalists would be able to corroborate or directly challenge. In practice, 'public' events are measurable and countable events; and White's counting and measuring has sometimes been played down in published versions of his work. (Particularly unfortunate is the omission from some editions of *Selborne* of his digest of the baptisms and burials in the parish for the sixty years from 1720 to 1780. A head for figures was another aspect of this many-sided man.)[1] But his quantitative awareness, above all, is what counterbalances his Selborne prejudice. Keeping the sort of records he kept, and comparing his garden with that at Bramshott Park, he has to admit that 'sandy soil is better for garden crops than chalky'. He describes the largest pond within Woolmer Forest as 'a vast lake for this part of the world', but he enables us, his readers, to make up our own minds about its vastness. Its total circumference is 2,646 yards, he says, or almost a mile and a half:

> The length of the north-west and opposite side is about 704 yards, and the breadth of the south-west end about 456 yards. This measurement, which I caused to be made with good

> exactness, gives an area of about sixty six acres, exclusive of a large irregular arm at the north-east corner, which we did not take into the reckoning.[2]

He had what some later editors have viewed as a mania for measuring. He measured the depths of wells and the heights of hills as well as he could, and he ascertained the output of Wellhead, the 'perennial spring', exactly.[3] He recorded the numbers of cabbage plants, sometimes even of cucumber seeds, he planted, and he measured the quantities of his orchard and garden crops. He weighed the tortoise each spring and autumn, in the scales at John Burbey's shop, because the grocer's scales were more convenient for the purpose or perhaps from pure friendship; and when he clipped his dog, Rover, he weighed the hair he had cut off. He counted the trees in a small wood he owned opposite the Long Lithe: 'There are 99 considerable trees in Sparrow's hanger; 94 beeches, 3 ashes, & two oaks: there are also three large oaks in the pasture-field adjoining.'[4] He became an inveterate measurer of the heights and girths of trees, and recorded these where possible in relation to their ages.

To the present-day reader, the counting of trees of various species in a wood may seem quite *un*extraordinary; and there was method in much of White's quantitative 'idiosyncrasy'. The minute calculations included in the paper on rush lights (DB xxvi) reflect the importance of rush lights in the lives of 'the poor', and his own detailed gardening records reflect, for one thing, the economic importance for him of garden produce; but these more obvious examples of his counting and measuring do not show us a tendency extraneous to the rest of his research. He went to lengths to keep it public and objective – or such as other naturalists could criticise and use.

Selborne itself illustrates the point. The non-naturalist may find the book not difficult but perhaps surprisingly plain and businesslike. White attributes feelings to his animal subjects, and was himself a sympathetic man, but 'sentiment' in the sense of sentimentality is almost never a feature of his writings.[5] *Selborne* is not a 'pastoral dream', and – despite a received conception which must have influenced great numbers of people to buy the book – it is not White's celebration of his native place. The 'diversity' within the Selborne parish gave it special advantages for the outdoor researcher, but complex 'economies' can be found anywhere, he implies. He deals with a range of species which occur in his parish for all or part of the year but all of which occur also in numerous other regions. He provides a detailed environmental background to his behavioural reports not in the process of eulogising a particular place but so that naturalists studying the same species in other, different conditions will be able to relate their findings constructively to his own. The reader who persists with the book finds not only its natural history contents but the person presenting these contents; but this happens in spite of a determined effort on White's part to be impersonal.[6]

Various of his poems were coloured by the Romanticism of the time, and Romanticism perhaps affected his antiquarian research; in imagination, he followed the intrepid voyagers to wild and distant parts of the world. But the publication of *The Natural History of Selborne* in January 1789, after years of preparation, is not once mentioned in the *Naturalist's Journal* for that year.[7] The twenty five volumes of the *Journal* can be called Selborne portraits, and in these White does refer to the 'high beauty' of his garden flowers or of a summer – or winter – landscape, but the same volumes contain virtually no allusions to his church duties. The entry for December 25, 1781, reads, 'A gardener in this village has lately cut several large cauliflowers, growing without any glasses. The boys are playing in their shirts', with an added note of some foreign news. Christmas Day was not then the elaborate business it became in the nineteenth century, but White, one hardly needs say, was entirely sincere in his Christianity. He was a better parson than many of his time; he rode to Faringdon in all weathers, during the years he officiated there, and he would allow himself no absence from home without having arranged a clerical substitute. But his journal entries – those of a scientific naturalist – were to be unaffected by moral judgments.[8]

Superstitious practices, and notably the 'cures' which belong in this category, are touched on repeatedly. The 'doctrine of longing', according to which congenital deformity in a baby was caused by some untoward thinking or wishing on the part of the mother to be, merely illustrated the gullibility of the 'good women' of the parish, he thought; but in recalling the ceremonies relating to a 'shrew ash', which were supposed to cure cattle or injured or deformed children, he represents them as both ridiculous and dangerous.[9] Even in 1751, in 'this enlightened age', two elderly women who were accused of witchcraft had drowned while being tried by ducking, at Tring, 'within twenty miles of the capital'.[10] He alludes in passing to the newspaper report of a 'wonderful method of curing cancers by means of toads', and points out the suspicious weakness in what the self-styled healer was saying. This person, who had 'cured herself' by the wonderful method, claimed that the secret had been passed on to her by a 'strange clergyman'. But why had it been a secret, White wants to know:

> Is it likely that this unknown gentleman should express so much tenderness for this single sufferer, and not feel any for the many thousands that daily languish under this terrible disorder? Would he not have made use of this invaluable nostrum for his own emolument; or, at least, by some means of publication or other, have found a method of making it public for the good of mankind?[11]

Superstition was inimical to science, but in ridiculing the shrew ash or those who believed in it, White was himself making a moral judgement. In telling us nothing about his own or the villagers' Christmas Day he was being too exclusive; this as much as ploughing or harvesting in the parish – or the placing of 'virgins' garlands' in the church, which he does describe in the *Antiquities*[12] – was conduct, or life and conversation. But by and large he does not edit the doings of his subjects, or edit them in this way. He restricts himself, on the strength of his moderate scepticism, but to what can be observed and measured in his animate and inanimate surroundings. At the same time, then, he restricts himself to aspects of those surroundings which make for collaboration and mutual monitoring. Popular report – untrained and unaffiliated contributors – could not be relied on; there was an unfortunate propensity in mankind towards 'deceiving and being deceived'.[13] But once Thomas Hoar had learned how to keep records, he could be given considerable responsibility. (Thomas sometimes manned the Selborne weather station unaided for weeks at a time, and it was he who was eyewitness to the umbrella behaviour of the spotted flycatchers.[14]) Where White knew that his brothers or friends were observing methodically and keeping records, as with the work on the stone curlews, which was instigated by White but carried out by Mr Woods of Chilgrove, he welcomed the information they could provide.

2 Scientific influences and examples

The research White passed on to his successors was incomplete; and if a personality was expressed in his working outlook, this outlook was also significantly shaped by cultural influences. Important among these, I have suggested, were a methodological tradition and a received psychology. The tradition gave a vital role to comparing and relating processes, and allowed research begun by one worker to be carried forward, if it survived criticism, by other and perhaps later workers.

Providing a clue to this tradition, David Allen remarks that the initial reception of *Selborne* was somewhat cool because 'the style of the writing was too Augustan to appeal immediately to a public accustomed to a diet of Romanticism'.[15] In both its style and its standards, *Selborne* was an old-fashioned, as well as a forward-looking, book. White did not forget his Latin poets (the outgoing but unpretentious bee cited on the original title page of *Selborne*, while it reminds us of the bee approved by Francis Bacon, is that of the poet Horace). John Ray was a constant companion; the late seventeenth century discussion of 'periodic birds' had sometimes been better informed than that taking place among White's own contemporaries. In addition, White owed much to Francis Bacon (1561–1626) himself, or rather to a 'modern' version of Bacon's conception of science. This was epitomised in the relevant work of George Berkeley (1685–1753) and David Hume (1711–1776), who are of importance here even if White did not himself read them. While also dealing with much wider questions, ~~logical and physiological~~, these thinkers show us the Baconian tradition taken to its logical conclusion. As they considered the scientist, or as he was often called, the 'natural philosopher', he examined and tested observables and the time and place relations between them, and gladly made use of mathematics, but he quite avoided so-called 'things in themselves'. The same scientist was interested in what happened in the phenomenal or natural world, and interested in how given events came about; but he made no appeal to 'occult' or 'ultimate' causes – material or spiritual.[16]

For John Locke, knowledge of any kind was comparative; to think was to try to discover how far given percepts or concepts 'agreed or disagreed'.[17] For Berkeley and Hume, the proper scientist searched for explanations, but where by 'explanations' he meant the *effective rules* which emerged as time and place relations were methodically tested. Using these rules, the scientist could offer qualified predictions such as we have referred to already; *in such and such circumstances, such and such further events can be expected.* The same theorists were not claiming that there were no occult or ultimate causes; rather, for the proper scientist, the search for such causes was unnecessary and unwise. A practical morality was consistent with this scientist's endeavours, and *vice versa*; as predictive, his explanations – his rules – helped the moral individual to act in the natural world with greater effectiveness and convenience. The scientist could arrive at his principal claims, however, without trying to transcend the observable and measurable aspects of that world.[18]

Even here, appropriate experimentation might be difficult to arrange. The scientific worker was restricted in various respects; his rules reflected what had happened but without also giving him insights into what 'must' happen. But as Newton had triumphantly shown, order generally prevailed in the natural world. And the proper scientist was not confined to *only his own* observing and measuring. Two people could agree as to the various objects they would describe as 'yellow', Locke had pointed out, even though neither of them knew what the other was experiencing when he perceived 'yellowness'.[19] In the words of George Berkeley, the natural world presented methodical observers with a 'universal language'; the rules ascertained by proper scientists would be the same or else compatible in all societies or parts of the globe.[20] In theory, then, numerous researchers, at a given time or in successive generations, could contribute jointly to the assembling and testing of data in connection with a given project. The method and subject matter of science were such that a 'more nice' observer, in White's phrase, could both criticise and add to a received body of facts.

This 'universality', or publicity, made important work possible, and provided a defence in practice. The researcher needed protecting against inadequately tested material – and against himself. The British thinkers *questioned* the 'authority of Reason'. Unaugmented intellect was weak and limited, they believed, and human beings as such were prone to self-deception; 'rational' accounts of the natural world might exemplify merely the trivial or perhaps vicious '*chimeras* of a man's own brain'.[21] Yet at this stage too a positive claim could be added. By concentrating on the proper materials of science, by concentrating on observables and measurables, the researcher could provide himself with worthwhile predictions, and could anticipate this fallibility. A self-indulgent rational*ising* was *out-manoeuvred*, as observational and public evidence was brought into play.[22] In certain circumstances, Locke says, even the direct observer can trust himself only if he has paused and carried out comparative checks; and as investigators, the empiricists all agree, we should avoid wherever possible the condition of the hermit-researcher – the self-satisfied recluse.

White is only rarely explicit as to his sources, and he was no doubt unaware of some of them, but several of his remarks on superstition and self-deception, or on the importance of remaining available to the surprises nature can spring on us, could almost have been copied from one or other of the tough-minded 'empiricist' writers. Whatever the relevance of these writers to the laboratory practice of their time, for several decades there were novelists, poets, historians and political theorists who thought of themselves as representatives of an empiricist 'spirit of accuracy'. A 'modern' Baconian intellectual atmosphere could be experienced during White's young manhood.

Solipsism was inimical to worthwhile science, and as White says himself: 'Ingenious men will readily advance plausible arguments to support whatever theory they shall chuse to maintain'.[23] White concerned himself with the 'improvement' of the Selborne parish, we saw, and voluntarily reinforced his links with the other residents. In a sense, the parish was a miniature, alternative world; the 'great lake', the hills and farms, the wild animals of the parish and indeed White himself, were all to scale. A *closeness* to the parish and its inhabitants might have quite subverted his attempts at objective examination. But – at his scientific level – White too made an ally of 'publicity'. Using public evidence, he out-manoeuvred himself. He could record violent events, even in the Selborne neighbourhood; a disabled hawk in a farmyard, which could not retreat before the angry mother hens, was set upon and torn to pieces by them, and male field crickets, though normally satisfied to merely 'sing' their proprietorship and defiance, will 'fight fiercely', he says, if confined together in a crevice.[24] This was observational and public material; other naturalists might be looking over one's shoulder, literally or otherwise; he could hardly avoid reporting the violence of the hens or crickets, therefore. But he recommends exchange and mutual monitoring. As a naturalist, he *calls for* 'candour' and collaboration.

The out-manoeuvering did not always work, as we saw while reviewing his contributions to the hibernation/migration debate; but generally, he rose to the challenge of accuracy and publicity. The paraphrase of Scopoli he uses, on the importance of first-hand and 'partial' reports, could imply merely that the findings of the naturalists concerned would be bricks added to a wall (when enough local monographs are fitted together, the result will be a 'universal correct natural history'); but he gave the passage an additional meaning. If the partial, or local, naturalists worked to the right common standard – specifically, if they were outdoor and behavioural in emphasis – they could contribute to an authenticating parliament of natural history. Some of the data White was himself offered by others was, as data, in a class with the reports of magical cures. The peer-examination of the evidence gathered by the 'partial' workers, as he envisaged this, would be as important as the gathering of the evidence in the first place.[25]

Observables and measurables allow for collaboration, and precise experimentation. John Ray, naturalist, was never far from White's conscious mind. White could describe 'physico-theology' as 'worthy the attention of the wisest man', but he made no serious contribution to physico-theology. (In practice, he did not accept that trying to construct the character of the 'Author of Nature' from a study of the living world was part of the naturalist's task.) But, in one connection or another, Ray's name occurs more than thirty times in the course of *Selborne* alone. Theology notwithstanding, and assisted by Willughby until that patron's early death, Ray had worked indefatigably on plants, insects, birds, fishes and 'animals with four legs', discovering, sorting and cataloguing. A passage such as the following, from *The Wisdom of God in the Creation*, perhaps links Ray with White directly: 'Some reproach methinks it is to Learned Men that there should be so many Animals still in the World whose outward shape is not yet taken notice of or described, much less their ways of Generation, Food, Manners, Uses observed.'[26] (If the program suggested here had been widely accepted, 'description' would have been merely a branch of nature study, and perhaps merely a part of the classification process.) White's own behavioural studies led him into disagreements with Ray, but his respect for this extraordinary predecessor – he 'conveys some precise idea in every term or word' – was never to fade.[27] Ray was sorry to see 'so little Account made of real Experimental Philosophy in this University, and that those ingenious Sciences of the Mathematics, are so much neglected by us'.[28] He was speaking at Cambridge, but might have been referring to any of the handful of universities existing in Great Britain at the turn of the seventeenth century. Ray eventually concentrated on the botanical taxonomy with which his own researches had begun; his master-work was his *Historia Plantarum Generalis* (1686–1704); but what he wanted in naturalists of any kind, what as he says they 'owed their Creator', was the sort of methodological exactness which was beginning to dominate the 'physical' sciences in his time.

Seven and eight decades later, and in his adult years, White was regularly in touch with people who were, at least, experimentally-minded. He paid visits to Philip Miller's nursery, where Miller had conducted some of the first research on insects as pollinators. While staying in London, or rather South Lambeth, he attended meetings of the Royal and Antiquarian Societies, although he did not read his hirundine papers before the Royal Society himself, and – unlike his brother Thomas – was never a fellow of the Royal Society. Thomas and Benjamin were hobbyists compared with Gilbert White,[29] but at Benjamin's shop and publishing business he could examine new natural history books – including new books from abroad – and meet other naturalists. We have no record of conversations conducted at the Fleet Street bookshop, but his connection with Thomas Pennant, and through Pennant the Hon. Daines Barrington, began there. It was perhaps by way of Pennant that he met Joseph Banks (1743–1820), the plantsman, traveller and – eventually, as Sir Joseph – Royal Society president. During the late 1760s, a formative time for White, he talked with the young Banks more than once at the latter's London home. There seem to have been no later meetings between the two, and despite White's hopes, Banks did not visit Selborne, but White followed the subsequent adventures of this brilliant amateur with especial interest.

White knew and corresponded with John Lightfoot, the Oxford botanist and friend of Banks. Through the mediumship of John White, and by letter, he once quizzed Linnaeus himself, and passed on to him information he, Linnaeus, made use of. Again as a young man, Gilbert enjoyed what he was later to call a 'most valuable friendship' with Stephen Hales, the physiologist, inventor and cleric mentioned earlier. Hales was at that time the incumbent at Faringdon; he was a non-resident, but he spent part of each summer there from 1722 to 1741.

Hales had been a keeper of phenological records, and his *Vegetable Staticks* (1727) is quoted respectfully in both *Selborne* and the *Naturalist's Journal*. In White's solving of the mystery of the 'little upland ponds', he was assisted by what he found in Hales's writings, and he gladly recalled this gentleman's conversation decades later for the benefit of Robert Marsham. Hales was unequivocally experimental; experiment meant 'analysing nature', he said, and should depend on 'number, weight and measure'.[30] Some of the work touched on incidentally in *Vegetable Staticks* would have appalled – rather than encouraged – Gilbert White. Hales was an unabashed vivisectionist, and perhaps eased his conscience as he thought of animals and plants as machine-like. White responds to vivisection too by not once acknowledging the practice, and by offering us quite other activities and insights; but he could apply himself with resolution to the dissecting of dead specimens – and he enjoyed technical contrivances.[31] Details of the pre-fabricated hot-house, and of a mobile fire escape he had seen demonstrated in a London street, appear in the *Journal*, as does a careful, eye-witness description of a flight by the balloonist, Jean Blanchard.[32]

White's connection with Thomas Barker, lastly, the brother in law who was an authority on meteorology, and who had been known to him since they were both children, may have been more important than it at first seems.[33] Barker was a link with those in eighteenth century Britain who were taking an intelligent interest in records. His quantitative weather summaries were being published annually in the *Philosophical Transactions*, and statistics relating to weather, human populations, and most impressively perhaps, epidemic diseases, were beginning to be used constructively during the second half of the century. Correlations between epidemics and local environmental factors were already being explored; in areas bordering on 'natural history', rudimentary forms of what would now be termed epidemiological analysis were already in use. The 'arithmetic' doctors were often under attack; but their expressed aim was to put the assessment of medical diagnosis and treatment, for instance in the realm of public health, on a quantitative and experimental footing.[34]

White was a supporter of this eighteenth century British 'numerical enterprise', as it has been called. He was required to provide human census and mortality figures for the Selborne parish, for a diocesan survey, in 1783. He tendered exact material covering a period of sixty years, giving average numbers of baptisms and burials for each sex in each decade and adding marriages from 1761 to 1780. Baptisms exceeded burials steadily, he showed, and 'chances for life' in males and females were equal. Gratuitously, as some have thought, he included these tables in *Selborne,* as a footnote to the fifth 'letter to Pennant'. In the first two 'letters to Barrington' he provides (1) lists of carefully ascertained chronological and comparative facts concerning migratory birds, and (2) several precise, quantitative schedules relating to bird song. Just such programmatic, factual material as this was what he hoped – largely in vain – to receive from other observers.

White was himself experimentally-minded; he did not take received wisdom, unsupervised observation, or even his own first impressions, on trust. Like that of the medical statisticians, his 'comparative analysing' was a sort of dissection, and was often itself a testing of hypotheses. But, while he understood what the other naturalists of his acquaintance were doing, they showed little understanding of what he was doing. He could write as if attempts to examine 'life and conversation' were already usual among natural historians; he was trying to encourage the behavioural studies he believed in.[35] He was often indebted to his village neighbours, but in cases of more advanced collaboration, he could exaggerate the contributions made by colleagues. In the *Natural History*, he on occasion almost invents colleagues, with a consequent playing down of his own original input. John Woods on the habits of the stone curlews in fact added little to the information White was already gaining unaided, but in summarising 'Woods's report' in the relevant *Selborne* paper, he was both advocating and giving an example of record-keeping and co-operative observing.[36] In practice, and playing off his own composite records one against another, he was often operating as a one-man team of investigators – and a one-man Royal Society.

Ray's classification was being superseded, and his behavioural injunctions were either forgotten or not taken seriously by most of his eighteenth century successors, but White could view behavioural research as indispensable to sound zoology or botany. Gilbert White is sometimes as 'puzzling' as the situations he himself tried to unravel – and if we cannot see how what he did was possible, we are inclined to under-rate what he did. (A member of the crew of the Kon-Tiki, a marine biologist, comes to mind. Shown an archaic 'snake-mackerel' which had landed on the raft during the night, he said: 'No such fish exists'.) White's attachment to behaviour is not quite inexplicable, it can be shown. But in the rest of this chapter I want to look at his meteorological and geological work, work in which he was generally in sympathy with the other scientists of his age. That climates and soils could and should be quantitatively examined was fairly obvious (although with soils and rocks, there were still obstacles to their thorough-going investigation), and in varying degrees both were features even of the county 'natural histories'. Only, for White, the true importance even of meteorology and geology lay in their relevance to the study of 'modes of life'.

3 The weather as natural history

Selborne, we can say, was written and re-written between 1770 and the middle of 1788. A collection of nature essays might have been produced using only, or little more than, the letters to Pennant and Barrington, and when as the years passed White's friends urged him to get the book finished, they were no doubt thinking of the descriptions of events at Selborne they received from him themselves in the form of letters. But the naturalist had his own ambitions for the work he at last committed to the press; his, it has been said, was 'the true humility which knew its own worth'.[37] The original, dated letters, though they were often altered, were not entirely re-cast; but to these were added the first nine papers of the 'Thomas Pennant' series and the last eleven papers of the 'Daines Barrington' series. The first are largely concerned with soils, rocks and water supplies; the second deal in part with climatic tendencies and exceptions.

'The weather of a district is undoubtedly part of it's natural history', White states,[38] and recording and questioning the weather gave his 'mania for measuring' special scope. In the *Journal*, he provides notes on the filling and management of barometers and warns us against mercury adulterated with lead (the latter has a lower 'specific weight' than mercury). He stresses 'the futility of marking the plates of barometers with the words – *fair, changeable*, &c., instead of *inches, & tenths*: since by means of different elevations they are very poor directions, & have little reference to the weather.'[39] If we are to be more than mere dabblers, thermometers, again, must be able to register extremes. One of his own thermometers proved unequal to the winter of 1784–5; 'when the cold became intense, & our remarks interesting, the mercury went all into the ball, & the instrument was of no service'.[40]

In the *Journal*, the daily meteorological records form an almost unbroken sequence for the years 1768 to 1793, though they include measured rainfall only from 1779 onwards. One of the very few harsh remarks to appear in White's notebooks or anywhere else in his writings, it has been pointed out, was the result of his not being able to provide a weather reading. In 1790, in June, while he was on a visit to South Lambeth, shade temperatures were recorded there of 87° and 89°F, but when he returned to Selborne he could not compare these with temperatures taken in his own garden on the same day. As he records instead, 'Thomas forgot to look in time'.[41]

When the less methodical notes in the *Garden Calendar* are added to the weather records in the *Journal*, the result is a meteorological account of the Selborne district covering some forty years. Early records, or those which seem reliable, have been invaluable to later students of climatic trends. White's precise rainfall data cover only a relatively short period – where the 'very intelligent gentleman' referred to in TP v, Thomas Barker, had kept exact records for more than forty years. But White does not give figures alone; he is exceptional for the variety and intelligence of his other *Journal* entries. In assessing the weather of an earlier era, the climatologist needs comprehensive descriptions of seasonal effects, and that White was a careful botanist, for instance, must be of interest to the present-day student of climates.

White himself sets his rainfall figures against those offered by Thomas Barker in Rutland, or Thomas White at South Lambeth or Henry White at Fyfield near Andover. After the mid-1770s, Selborne returned to its more usual, variable rainfall: between 1780 and 1792 inclusive the average annual rainfall at The Wakes was 36.43 inches, but where the highest year (1782) was 50.26, and the lowest (1788) was 22.50 inches.[42]

At Selborne, during White's adult lifetime, shade temperatures rarely exceeded 80°F. Selborne was a relatively cool spot, and in the second half of the eighteenth century, Britain as a whole was still in the final stages of the so-called Little Ice Age. (It affected Great Britain for more than four centuries, it seems, beginning early in the fifteenth century.) Among the composite descriptions of winter weather to be found in *Selborne* and the *Journal*, White's notes on the frost and snow of January 1776 stand out. On January 14th, 1776, after the first heavy snow falls, 'The narrow lanes are full of snow in some places, which is driven into the most romantic, & grotesque shapes. The road-waggons are obliged to stop, & the stage-coaches are much embarrassed.'[43] Hares were driven by hunger to enter the village gardens during this 'rugged' weather, and 'Lambs fall, & are frozen to the ground'. Rooks and skylarks frequented the yards close to buildings and small birds were easily caught by cats.[44] It was difficult for visitors to get into the Selborne parish in winter, but when necessary, even in extreme conditions, Gilbert White would get out; and on the 22nd of January he 'had occasion to go to London'. He could enlarge his observations, therefore. He passed through a wild and unrecognisable – a 'Laplandian' – countryside, and on reaching the capital, found the streets 'deeply bedded in snow'; the horses and carriages were running, he says, but running without the least noise. By January 31st, when snow had lain on the London rooftops for twenty-five days, the temperature fell to 6°F; the Thames was frozen over both above and below the Bridge. Crowds ran about on the ice; but Selborne was still undergoing its exceptional weather, White can add. There on the morning of the 31st, 'just before sunrise, with rime on the trees and on the tube of the glass', the thermometer fell to zero (or thirty degrees below freezing) on the Fahrenheit scale.[45] On this occasion, clearly, Thomas Hoar had been especially watchful.

When White returned to the parish, in March 1776 – among the matters occupying him 'in town' had been work on the fourth edition of Pennant's *Zoology* – he found that the effects of this snow and frost at Selborne had been relatively slight. Both the winter wheat and the turnips had emerged looking well: they had been 'blanketed and protected' by the snow. And his own laurels and laurustines were less damaged than they might have been, because Thomas, acting on instructions, had shaken the snow from their leaves and branches.

White began using this snow-shaking specific after his experiences in the extreme frost of 1768,[46] and it was to entirely fail him only once. This was during the winter 1784–5.

On December 10th 1784, his Dolland thermometer fell to one degree below zero, at Selborne. The 'strange severity' of this month had made him desirous to know 'what degree of cold there might be in such an exalted and near situation as Newton', on the top and to the other side of Selborne Down. (Newton Valence was, by his reckoning, three hundred feet higher than his Selborne house, and was 'near' because of the steepness of the north-east side of the Down; to compare temperatures taken in the two villages was to compare temperatures in more or less the same place but at different altitudes.[47]) He had sent a message to the rector of Newton, therefore, urging him to 'hang out his thermometer'. He assumed that the Newton temperatures would be even lower than his own; surely temperatures fell as one went higher; but to his surprise, on the 10th of December, the Newton thermometer had registered 17°F, and the next morning, when the temperature at Selborne was 10°F, that at Newton had reached 22°F. He reacted to this unexpected 'comparative local cold' in the manner of a true researcher. He sent his own thermometer up to Newton Valence, to be placed beside that of the Newton observers. The two instruments 'accorded exactly'. During one night at least, then, the temperature at Selborne had been eighteen degrees lower than that at the higher and more exposed village.

Subsequent observations bore out this difference. Despite the snow-shaking, White's laurustines, bays, ilexes, arbutuses, cypresses and even his Portugal laurels were all 'scorched up', during this bitter period, whereas the same or similar species at Newton 'did not lose a leaf'.[48] His walnut tree was so damaged that the following year he was quite without walnuts for pickling – but the friends at Newton sent him some! Beyond his garden, the frost of this winter killed 'all the furze and most of the ivy', and stripped some hollies of all their leaves. The cold weather came early, White says, and yet, from its effects, it exceeded any since the winter of 1739–40.[49]

W.B. Turrill might have been thinking of Selborne when he wrote of the British Isles that, 'the micro-climatic, micro-edaphic, and micro-biotic factors vary within small areas, and comparisons are thus easily made *in situ* and under field conditions'. (Useful 'comparative observations and experiments can be made to merge or at least can be carried out together, and all without waste of time in traversing great distances.')[50] White, though his readings showed clearly that at Selborne and Newton Valence the simple relationship between temperature and height above sea level did not obtain, could think of no explanation for the 'reverse' just described. (He was unaware of the 'cold slides' and 'cold hollows' because of which, in certain conditions, the temperatures at the bottom of a hill may be kept at a relatively low level.[51]) He was sometimes surprised not only by given meteorological events but by the explanations of these events others had provided. In June 1783, he makes this note:

> Dr Derham says, that *all* cold summers are wet summers: & the reason he gives is that rain is the effect not the cause of cold. But with all due deference to this great Philosopher, I think he should have said, that *most* cold summers are dry; as for example, this very summer hitherto: & in the summer of 1765 the weather was very dry, & very cool.[52]

He reached positive conclusions, though, in several enquiries arising from warm summer weather. He made careful studies, we said, of periods of drought. He was intrigued by the 'upland ponds', two of which, on the top of Selborne Down, were near Newton Valence. These ponds – they were 'dew ponds', a name adopted at a later date in part because of what White had written about them – were small and shallow, but they remained viable sources of water even during periods without rain, and even where large numbers of sheep and cattle were drinking from them. In an earlier chapter, I mentioned these ponds in connection with the year 1781, a year of low rainfall which followed a year of low rainfall. Temperatures seldom rose above 70° F, at Selborne during 1781, but White describes the middle of the year as 'a severe hot dry summer'. Clearly, the drought and the 'heat' combined took their toll of the region. To quote one of the relevant *Journal* entries at greater length:

> The greens of turnips wither, & look rusty. The distress in these parts for want of water is very uncommon. The well at Grange-farm is dry; & so are many in the villages around: & even the well at *Old-Place* in the parish of E. Tisted, tho' 270 feet, or 45 fathoms deep, will not afford water for brewing. All the while the little pond on Selborne down has still some water, tho' it is very low: & the little pond just over the hedge in Newton great farm abates but little. The ponds in the vales are now dry a third time . . . The last wet month was Decem^{r} 1779 . . . since which the quantity of water has been very little.[53]

In White's locality, these upland ponds were to be found on chalk hills, which enabled him to rule out one possibility at the beginning of his enquiry; the ponds were not fed by 'secret springs'. They did not overflow in a *wet* year, and anyway, water sank through chalk, forming a 'level' and producing springs only when it reached a less permeable soil or rock.[54] As I implied earlier, he found the essential clue in the way moisture would collect on and under foliage at night, in the parish; the trees which over-hung certain of the ponds contributed to their replenishment. He then remembered the claim of Stephen Hales, that 'the most dew falls on a surface already moist'. The integrating step he took himself. In addition to the effects of adjacent foliage, the *surface water* of the ponds acts as an 'alembic', he decides. The surface water, 'by it's coolness, is enabled to assimilate to itself a large quantity of moisture nightly by condensation'.[55]

In a letter to one of his nephews, he describes an experiment which, he says, would provide valuable evidence for or against this eventual theory. Suspend two identical sponges over an exposed pond, one close to the surface and the other a yard or two higher up, and leave them there throughout a misty night. If the lower sponge contains more water in the morning than the higher sponge (the control), the case for 'water condensed by water' will have been largely made out.[56]

White seems not to have performed this experiment himself, although it again illustrates his special acuity. But, in his 'moist' local climate, he gave frequent attention to condensation – and evaporation.

The importance of evaporation was brought home to him as he read 'Watson's Chemical Essays', and confirmed from personal observation that there was water vapour in the air, in his region, even during the hottest and 'driest' weather.[57] His talk of cold descending from a clear sky at night, or of a thaw beginning under ground as warmth 'gets the upper hand' of cold, must now seem merely quaint, but his understanding of a water cycle – or rather, a regular redistribution of water – operating in his locality, which appears when various scattered notes are brought together, is perceptive and seems to be original. Moisture evaporated from lowland wooded areas only gradually, he thought, and Selborne was well provided with such areas. Even during periods of drought, moisture from these woodlands would spread across the higher and more open parts of the parish in the form of night-time mists, and because of 'dripping', would be returned to the ground in places which would otherwise have been particularly dry.

As happened several times during 1781, Selborne might be

without rain while rain was falling on one or more of the surrounding parishes. Thunder storms often by-passed the Selborne parish, White explains. Storms usually arose to the south or south-west of Selborne, and it seemed to the naturalist that Selborne Down and Noar Hill, and behind these Barnet, Butser Hill and Portsdown, diverted and divided the storms as they were moving inland.[58]

The 'sheltering' effect of these hills could be a mixed blessing. But in June 1784, the villagers would have been glad of this effect. The great hail storm mentioned earlier, a prodigy in all respects, moved towards the village from the north, and met with no local barriers. White's precise description of this hail storm begins:

> . . . on June 5th, 1784, the thermometer in the morning being at 64, and at noon at 70, the barometer at 29 – six tenths one-half, and the wind north, I observed a blue mist, smelling strongly of sulphur, hanging along our sloping woods, and seeming to indicate that thunder was at hand.

Piecing the story together several days later, he goes on:

> At about a quarter after two the storm began in the parish of *Hartley* . . . from thence it came over *Norton-farm*, and so to *Grange-farm*, both in this parish. It began with vast drops of rain, which were soon succeeded by round hail, and then by convex pieces of ice, which measured three inches in girth. Had it been as extensive as it was violent, and of any continuance (for it was very short), it must have ravaged all the neighbourhood.[59]

The roaring of the hail as it approached was 'truly tremendous', and rocks weighing two hundredweight were washed out of the sides of one of the hollow lanes. The damage to White's garden glass was not as extreme as might be gathered from the *Selborne* letter, but when the frames were mended, early in the following year, the glazier's bill as noted in the *Journal* came to two pounds, five shillings and ten pence.[60]

In a lighter mood, he tells us that his part of the parish, with its abrupt hillsides, is 'another *Anathoth*, a place of *responses* or echoes'. One of the local echoes could 'repeat ten syllables most articulately'. This aroused the tireless inquirer; he asked what atmospheric conditions most facilitate the transmission of echoing sounds. A 'dull, heavy, moist air deadens and clogs the sound', and hot sunshine or a 'ruffling wind' largely defeat the efforts of 'echo'; but she performs her task admirably, it seems, on 'a still, clear, dewy evening', when the air is 'most elastic'.[61]

He leaves an interesting record of the exceptional occurrences of 1783, a warm year with persistent thunder storms in 'different counties'. During mid-summer 1783 – most notably, at Selborne, from June 23rd to July 20th – a lowering 'smoky fog' built up, so that the sun could provide merely a dull, 'ferruginous' light even at mid-day.[62] The whole affair was viewed by many of the villagers with 'superstitious awe', needless to say, but White draws our attention to the fact that, during this four week period, a volcano 'sprung out of the sea' off the coast of Norway, and parts of Sicily and Calabria were convulsed by earthquakes.[63]

He is hinting at a connection between these various events; the volcano might have been the cause of the smoky fog. He is at his most confident when working from his own observations; and yet, he knew, Selborne weather could not be considered constructively without the help of data from other regions. One cause of his preoccupation with travel books was that 'I love to study and compare climates'.[64] What Dr Johnson says about the 'black spring' of 1771, in his account of his Scottish tour, agrees closely with White's own records for the first months of that year, the latter points out.[65] He gathered meteorological material from newspapers – by the 1780s, he was receiving three or four newspapers a week at Selborne – or from the more distant of his correspondents; this is often among the items added to the *Journal* entries retrospectively. Some Selborne weather might be peculiar to it, but on the day of the great hail storm, a much more devastating hail storm occurred near Taunton in Somerset. Mountain snow melted in Hungary during 1781, the 'severe' drought year, and vineyards were frozen in Austria in June 1784, a cold summer month in Hampshire. John White's book, his natural history of Gibraltar, should include a comparative review of the climates of Gibraltar and southern Britain, Gilbert suggests on several occasions. In 1775, he asks, 'Should you not produce in your work a short comparative table of weather at Gib., Selborne, and N. America', going on to tell his brother where he can find the American records in print.[66] White's own most persistent travelling, in his young manhood, did not take him beyond Cambridgeshire and the English Midlands and West Country, and Pennant, working 'to the north' of him, and John White both proved much less capable than he had hoped of making use of their geographical and climatic situations relative to his own. But in the opening paragraph of *Selborne*, Gilbert White locates his parish not merely within southern England but on the globe itself. Selborne, he tells us there, lies half way between Alton and Petersfield, about fifty miles south west of London, and in latitude 51.

4 Varying soils and puzzling fossils

White's internationalism can surprise us. In a letter to John White, he notes that 'Lin: mentions 12 species of Hirundines, & Brisson (a French modern ornithol:) describes 17 species'.[67] In a *Selborne* letter, he says: 'Linnaeus ranges plants geographically; the palms inhabit the tropics, grasses the temperate zones, and mosses and lichens the polar circles; no doubt animals may be classed in the same manner with propriety.'[68] In the *Journal* – a home production but with its brief but frequently added items from other places – he was portraying and 'questioning' his parish and its inhabitants, and was producing a model of work in progress. While concentrating on his own patch, the researcher should not be a mere miniaturist; he should be intimate with a given locality but should, in certain senses, also be a man of the world.

The charting of climates was already taken seriously, by White's time. (One feature of Cook's circumnavigation and first voyage to the South Seas, from 1768 to 1771, was careful meteorological description and measurement.) He could relate his own 'climatic' work fairly easily to wider and more comprehensive studies. Systematic geology, alternatively, was still in its infancy in the mid-eighteenth century. The importance White attached to soils and terrains is clear from his placing his geological papers first in the *Selborne* collection, but in the case of geology, the more comprehensive charts had barely begun to be drawn.[69]

The early history of geological formations remained largely or entirely mysterious; a reverie, a half humorous speculation, mentioned by White, can illustrate the point. While riding along the Sussex Downs, he has sometimes asked himself, he says, why these smoothly rounded hills are so different in appearance from some other examples of high ground. Is their present form the result of a 'leavening fermentation', which raised them at some earlier period above the clay of the Weald? He seems to have been thinking here of yeast, or pans of rising dough.[70]

But this was a reverie only; and although White was active and inquiring virtually until the time of his death (his only serious inconvenience in old age was a certain periodic deafness), he had more than enough to occupy him in describing the present geological character of his region. With little other than his own persistent, thoughtful examination to go on, he distinguished the various areas of surface rock and soil in the parish, as one major project, and effectively mapped the extent of each. As he correctly implies, these areas, or the areas other than the chalk itself, lie roughly parallel to the north eastern slopes of Selborne Down and Noar Hill, in bands of varying widths. Along with the valley in which the Selborne stream runs, and allowing that the eighteenth century parish included part of Woolmer Forest (unlike the smaller present-day parish), they more than anything else account for the 'diversified and uneven' country White refers to in numerous reports.

Roughly, the parish was five miles in length, from west to east, and varied from one to four miles in width. In conducting us across the topographical bands, he starts with the chalk hills themselves. They are the highest parts of the parish and lie on its south-western border, with Selborne Down rising to some six hundred and fifty feet above sea level.[71] Moving north-eastwards – keeping to the sequence as White gives it in the early *Selborne* papers, but with assistance from Walter Johnson – we come next to the narrow strip of chalk marl at the foot of the hills, where White's own ground lay. In this strip both tributaries of the Selborne stream rise, the Bourne stream and the 'perennial' Wellhead. Beyond it, 'on the other side of the village street', lies an equally narrow strip of sandy clay; this is the 'black malm', described by White as a forward gardening soil. The black malm thinly covers a 'freestone' or malmstone, fourthly. The freestone is the foundadtion on which the village street is built; it is encountered first as rock, but as we continue north-eastwards, and in its disintegrated condition, it is the main ingredient of the 'white malm' and 'white land'. The white land is a freestone rubble with some admixture of sand, according to Johnson; hops thrive on it, White says, and 'root deep into the freestone beneath it'; it extends as far as, for instance, Temple Farm.[72]

Beyond the white land lies the gault or clay, fifthly. Priory Farm is situated on the gault, and it is exposed in the valley of the Selborne stream: it – White refers to it as 'strong land' – extends as far as Blackmoor. Beyond it, a sandy loam marks the transition from the gault to the sand, the boundary running out towards Woolmer Pond and then, south of Woolmer Pond, passing through Greatham. The 'sand country' completes the sequence. Oakhanger hamlet and Shortheath Common, where White would watch the sand martins, both lie within it, and it includes much of the 'sandy, treeless' Woolmer Forest. The Forest extends beyond the north-eastern edge of the parish, and although it is boggy in patches, some of it, again, is rocky and is relatively high ground.[73]

In White's geological explorations too, the range of his local knowledge stood him in good stead. The locations of plant and animal populations, wild or artificially developed, and the positions in the parish of the various 'geological bands', frequently coincided. (If he had reduced his work on each to diagrams on transparencies, the diagrams would often have coincided). The farming of the parish was innocent of large-scale amalgamation, and as he both observed and took part in the local gardening and farming, his knowledge of local soils was rendered more coherent. The gault clay allows for corn-growing and also brick-making, he notes. The local *walls and buildings* coroberrated various of his geological insights. In places, the sand of Woolmer Forest was a hard, 'crystalline' sandstone, it is identified by Johnson as Folkstone Beds; used as it was quarried, it made a rough but permanent building material. Another 'stubborn and rugged' stone was the 'blue rag', found in thin layers within the freestone. It was valued as flooring for stables, and because of its durability, it was used by White himself in the wall of the ha-ha.[74] The freestone, though it may not contain lime, is largely constituted of carbonate of lime, according to Johnson; White describes it as 'approaching a limestone', and thinks the presence of beech trees on or above it must be of significance.[75] (The Short Lithe, a freestone outcrop, is today largely covered with beech trees.) The freestone did have a bed to it, despite its name; if it was not cut and laid correctly, it would flake and break up when exposed to the weather; but some of the village houses were (and are) constructed of it. It was in request for hearth stones and the beds of ovens, and it worked admirably as a lining for lime kilns.[76] The main village street showed where the freestone was near the surface, and Selborne had grown up 'in the shelter' of Selborne Down. The fastidious White could have written a paper from hard-won experience on the geology of communication, communication, that is, by lane and road in his part of Hampshire.

White recognised, of course, that these rocks and soils were the slanting cross-sections of strata. One reason for the close interest he took in the wells of the village and the surrounding country was that they offered him geological data: well shafts are regular perpendicular borings, and with the help of what he learned from the wells and well-diggers, he could render his geology generally three dimensional. Within the village the wells were commonly a little over sixty feet deep, and at this depth they 'rarely failed'; after even the severest drought, four or five inches of rain would quite replenish them. These wells penetrated the freestone and reached a water table on a stratum of the ragstone or perhaps the gault itself, he thought. (The soil at

Ringmer in Sussex is a fertile 'brick loam', he says, with 'Clay under, which holds water like a dish.'[77]) When Timothy Turner, whose well had been only fifty seven feet deep, employed well-diggers to deepen it in January 1789, he was no doubt obliged to White for directions. Gilbert attended, anyway, and kept a record of progress. The diggers added nine feet to the well, and at one stage seemed to have been successful; 'they came today to a hard blue rag, & a little water'.[78] But they did not find the reliable further water supply they were looking for. Though White's own well still contained three feet of water, his conception of the local strata was too simple, we now know; at these depths, in the village, the impervious layers are intermittent.[79]

His 'three dimensional' picture is, nevertheless, deeper and wider than anything the village wells alone would have made possible. All the farmhouses and some of the scattered cottages of the locality used well water, and within a mile and a half of The Wakes there were wells three and four times as deep as his own. We do not hear of Gilbert White's descending into a well; but, assisted by his nephew Edmund, he measured the depths of many of these well shafts. The well at Goleigh Farm, on the chalk beyond Noar Hill, was more than 172 feet deep, according to his notes, and the 'deep & tremendous' well at Heards Farm, on the further side of Selborne Down but still on the chalk, was 250 feet to the bottom.[80]

White found valuable sources of geological fact in the 'rocky, hollow lanes' of the parish. These ancient cart-ways were worn down by weather and human use 'through the first stratum of our freestone, and partly through the second'. In some places they lay as much as eighteen feet below the surface of the adjacent fields.[81] He examined stone quarries and clay and sand pits; he once, at least, compared types of sand using a microscope.[82] Presumably he could think of the chalk downs, seriously or otherwise, as having somehow risen to their present size, because these 'immense calcarious masses' seemed relatively homogeneous (no springs break out in chalk). But he noted, and thought about, the extreme steepness of the north-eastern slopes of the downs. These escarpments, though usually well covered with turf or beech trees, he associated with falls of rock and soil. A notable rock-fall had taken place near the neighbouring Hawkley in 1764, when part of a north-eastern facing hill-side had collapsed. It left an exposed cliff about seventy feet in height and 'resembling the side of a chalk pit'.[83]

To return to the strata as strata, but to return indirectly, the 'fossil-shells' of the parish must not be passed over in silence, White says.[84] The paper in which he briefly reports on the fossils he has personal knowledge of, itself placed near the beginning of *Selborne*, must be properly acknowledged in this chapter. This paper again tells us about received information, or the lack of such information.

The first fossil specimen he mentions had been ploughed up in one of the chalky fields near the Down. A bi-valve, it is called by collectors a 'cock's comb', he says, and is rated a great curiosity: he names it as Linnaeus's *Mytilus crista-galli*, and it is shown in a large engraving in the early editions of *Selborne*. He had gone to great trouble to make this identification,[85] and in 'the museum at Leicester House', in London, he thought he had seen 'several of the shells themselves in high preservation'. The creatures concerned may be found alive in the Indian Ocean, he gathers, but he offers no suggestion as to how or why a fossilised specimen should be present in a field in Hampshire. If the cock's comb was a curiosity, ammonites were 'very common' fossils in his part of the country – then as now; apparently, the labourers found them just beneath the surface when the bostal path was being dug on Selborne Down. In a marl pit under the eastern slopes of the Down, he had seen ammonites of fourteen or sixteen inches in diameter, and in a chalk pit, at the north-west end of the Hanger, he had found 'large nautili'.

For later generations, fossils were to be the great indicators of the ages of strata, but no such understanding was available to White. The large ammonites, he says, consisted not of 'firm stone' but of a kind of hardened clay: they mouldered away when exposed to the rain, and 'seemed to be a very recent production'. The last sort of fossils – or true fossils – he refers to were recovered from time to time by the well-diggers. Large '*scallops* or *pectines*, having both shells deeply striated and furrowed alternately', they were found in 'the very thickest strata of our freestone, and at considerable depths'.[86] If some fossils were 'recent productions', these last examples, clearly, were exceptionally old.

Flints are a kind of fossil, but White had no reason to relate these to once-living creatures. If the chalk hills did not seem to him to be stratified, he could not have failed to register the layers of flints in the chalk, exposed in chalk cliffs or in the sides of chalk pits. But the strata he did recognise and distinguish can have been little more than strata, to him; and compared with those examining downlands such as he describes even a few generations after him, White – the master observer – would have viewed flints almost without seeing them. He includes and describes the fossils he was aware of, and he makes no explanatory reference to the biblical Flood; but his lack of any alternative to this explanation is plain. He knew the fossil specimens he recognised were 'casts', and he appreciated in the instance of the pectens that, although they were found in solid stone and in his Hampshire parish, the creatures of which they were casts had been aquatic. But, by subsequent standards, he knew little or nothing about *geological time*. He makes no suggestions as to the ages of the strata he uncovered. In connection with fossils and strata, he could make *no* significant comparisons.

5 The 'controverted Design'

The distance of the earth from the sun could be accurately calculated (this was one result of Cook's first voyage to the South Pacific), but prehistoric studies were still principally informed by myth in the second half of the eighteenth century. For many of White's contemporaries, there had been no prehistory in our own sense. Beyond or before modern history there was the history of Rome and ancient Greece, and the largely guessed-at histories of Egypt or Assyria; overlapping these, there was biblical history, which took one back, generation by generation, to the time of the Creation. With more or less conviction, a notion of the age of the created universe based on the biblical genealogies was still widely accepted; the universe had existed for a little more than 6,000 years.

The effects of this belief were profound, and 60,000 years, instead of 6,000, would have made little difference. Perhaps predictably, Daines Barrington was fanciful on the subject of fossils,[87] but early palaeontological discoveries could be misreported or otherwise imposed on, even by the representatives of Reason. Voltaire was among those who viewed many fossils as having no biological origin at all; bone-like fossils which did not match the bones of any living creature were merely stones which for some reason were bone-shaped. As is clear from an early edition of the *Encyclopedia Britannica* (1777–83), the response to 'gigantic' bones could still be equivocal. One (unnamed) author, after describing some gigantic bones, agrees that Sir Hans Sloane has dismissed them all as the bones of 'elephants and whales', but points out too that 'certain anatomists' have found many of them to be human bones.[88] By this date, some readers of the *Encyclopedia* would have rejected both Sir Hans Sloane and the giant humans; but even in the 1820s, the generally well-informed William Cobbett could be mystified by 'the lines of flints running parallel with each other horizontally along the chalk hills'.[89]

The 'Design' theory of the organisation of the natural world persisted, as long as this time scale of thousands of years persisted.

Religious orthodoxy was again the corroborator; and in its early forms the theory was pessimistic, on balance. The observable living world was an organisation, and was part of a hierarchy; the hierarchy extended from the natural to the supernatural realm, with human beings placed half way up, as supervisors of the natural. (William Borlase's remarks on the purpose of the physical differences between members of some animal species were not as extraordinary as they now seem.) Because of the Fall of the first humans, the earthly part of this totality was imperfect, and was ultimately doomed.[90] But the living species to be encountered on earth at any time were the original species; the Creation had been all-resolving, at least to this extent. If a bird, a fish, a horse bot or a bird's nest orchid seemed especially well adapted to its usual surroundings, this was because, at the Creation, it had been thus equipped and arranged. The woodpecker therefore has a long tongue, proper for searching out ants and other insects, John Ray had remarked. The same bird has short but very strong legs, with the toes standing two forwards and two backwards; this disposition 'Nature, or rather the Wisdom of the Creator, hath granted to woodpeckers, because it is very convenient for the climbing of trees'.[91]

We can find ambivalence regarding Design throughout the eighteenth century, but without an entirely revised time-scale for both animate and inanimate nature, even those trying to break out of this theory could make little meaningful progress. (Pictures and statues of animals from the 'earliest periods' could still be examined, after all, and seemed to represent plants and animals either identical to or very like our own.) Ray had himself been 'quite confounded' by fossils; they could be found in deep strata, and it was at least possible that some of them represented not only once-living creatures but species which could now be known to naturalists only in their fossilised forms.[92] Ray did not pursue his prevision here, because of the apparent implications. If these were facts,

> there follows such a train of consequences as seem to shock the Scriptural history of the novity of the World; at least they overthrow the opinion generally received, and not without good reason, among Divines and Philosophers that since the first Creation there have been no species of animals or vegetables lost, no new ones produced.[93]

White offers no opinion regarding 'extinct species', and has little to say directly to divines and philosophers; but he was alert to contemporary discussion. He was a student of John Ray, and was familiar with Robert Plot's county surveys. Plot, though he accepted that many fossils 'could not be' the remains of animate creatures, summarised this 'great Question now so much controverted in the World' as it was being debated in his (late seventeenth century) era.[94] White had no doubt about the organic origins of the 'fossil shells' he was referring to; and the adaptable animal he describes in his notes on instinct and 'ingenuity' can seem to bring a once and for all Design into question. On unfrequented islands even the large birds were often so unafraid of man that they could be killed or captured without difficulty. Confronted with the notion that Europe, Africa and the Americas had once been connected, White was quite nonplussed; but his '*congenerous species with divergent styles of life*' were inconsistent with 'Design' as it has just been referred to.[95] He, White, would not be hurried into absolute flexibility; the behaviour of the adaptable creature too, he could see, was structured and somehow directed. But he had begun to regard some even among inborn responses as theoretically alterable; in other circumstances they might have been different or might not have appeared at all.

The gloom pervading Ray's later writings is quite absent from those of Gilbert White. A new notion of 'nature' was, as it were, infiltrating the Design concept during White's adult years. The *natural world* showed 'capability of improvement', and if not the observable natural world then a personified 'Nature' was itself a creative source. In France, even by the middle years of the century, the admirable Maupertuis – drawing upon himself the vindictiveness of a Voltaire – was arguing for extinction and 'descendent mutation'.[96] During the same period, Linnaeus (the 'greatest living naturalist') continued to insist on the *im*mutability of species more or less dogmatically;[97] but by the end of the century, Lamarck and Erasmus Darwin were referring openly to contingent, evolving species and to millions of years of prehistory.[98] Soon a rational accommodating of the fossil evidence would be much more difficult to avoid.

The Design concept was still accepted unquestioningly by most of White's fellow Anglicans (even if, like White or for that matter John Ray, they viewed 'man' as to some extent self-determining). The early evolutionary skirmishing took place principally in France; during the eighteenth century, and though the progressive agriculturalists were internationally recognised, British biologists 'made virtually no contribution to evolutionary thinking'.[99] A dualism regarding Design is present in White's own writings; a designed and unchanging natural world is present in *Selborne* and the *Naturalist's Journal* along with fossils and the appetitive and adaptable animals. And yet, the story of White's adumbrating of later research does not end at this point. The 'dualism' seems not to have caused him serious self-doubt. In one of the *Selborne* papers, we remember, he wonders whether nature observes any 'mode or rule', but where this is meant merely to shock and stimulate his eighteenth century readers.

His remarks on plant sexuality can be noticed again here. He assumes that among plants intercourse between the sexes is central to a major means of reproduction, or is not merely the occasioning cause of this reproduction. As a botanist, this is to say, he quite rejects so-called 'preformation', the genetic and embryological principle accepted by Ray and his contemporaries and retained by the 'more modern' Linnaeus. Preformation was an offshoot of the Design theory. In any species, each generation replicated the previous generation because, according to the preformationists, not just the first representatives of that species but all subsequent representatives had, in a sense, been placed on earth 'at the Creation'. Reproduction added new members to a species reliably, because miniature animals or plants (and in a more complicated version, miniature but complete sets of the appropriate animal or plant components), were contributed by one or other of the parents during the sexual encounter. There was a sperm theory and an ovum theory; but in either case, the ready-made miniatures contained even more minute ready-mades within them, and so on. An infant, though it started to develop as a result of 'sexual conjunction', did not originate in that conjunction. A blending of the heritages of both parents was not part of the process. Most importantly – and except in the cases of 'monsters' – there was no allowance for unprecedented blends.

White distanced himself from this unlikely dogma. He noted individual differences, and he saw that the characters of both parents might appear in inter-bred farm animals. He saved flower and vegetable seeds selectively; he was curious about processes of fertilisation in plants, and as a serious gardener would have known that both 'sports' and plant hybrids can, in many instances, reproduce.[100] He nowhere argues against 'preformation'; he by-passes it, rather, along with the rest of the micro-biology of his time.[101] He assumes that insemination and ovulation are both contributors, in the production of new members of a species, but assumes also that these, as we would call them embryological, processes need not concern him.

His geological and meteorological surveys, identifying the diversified soils of the parish and plotting the varying moods of the weather, can be thought of as additions to larger geological and weather models, but they show us, equally, the conditions in which he could continue his behavioural enquiries. His paper on fossils is a quite unassimilated part of the *Natural History of Selborne*, even if we can see why he would not leave it out, but despite the factual and fragmentary nature of the book, this can be said of far fewer of its contents than might appear to the casual reader. Typically, White concentrates on the observable responses of animals and plants, and within this keeps to a number of 'theme topics'. And now some of the advantages which went with this typical stance become particularly apparent. The *time barrier*, though its presence can sometimes be felt in his work, did not seriously impede him. Dealing in behaviour he encountered flexibility almost inevitably, and behavioural responses can be recorded and compared *whether or not* one is aware of prehistoric time. Because of the material White concentrated on, he had what was in practice an antidote to the time barrier. The alterability of physical 'structure', in living things at large, could hardly be recognised so long as the world was thought to have existed for merely thousands of years, but present or past behavioural alteration – once a behavioural subject matter was accepted – could be contemplated quite easily. A new habit could become an established habit, within a regional population for instance, in the course of relatively few years, and adjustments of this sort could certainly be observed within the working lifetime of one natural historian.

White, though he evaded embryology, assumed that organised 'outdoor' research would contribute to the better understanding of living things. Indirectly, and within the concepts available to him, he emphasises the time factor. Changed behaviour becomes one of his theme topics; he was on the watch for it, plainly, and careful and laconic though he is, he finds it in an extensive range of creatures. He points to behavioural and ecological diversity within species – and within the larger generic groupings he refers to as 'congenerous species'. He offers little thorough-going explanation of this diversity. He does not envisage fundamental changes to the ecology of the Selborne parish. But it follows from many of his remarks that, if its ecology were to change, some of the living creatures to be found there, rather than dying out, would react constructively to this circumstance. Either they would stay where they were and re-adapt, or they would find, and move away to conditions at least resembling the conditions with which they were familiar.

Chapter 10

THE PROBLEM OF ADAPTATION

1 New departures and old beliefs

A nineteenth-century attempt not only to explore but to master nature had little counterpart in the work of Gilbert White. Various of his 'hints' received some endorsement in the century following his own, but his hopes would not be fulfilled by competitive or unco-ordinated naturalists, or by naturalists who, even if they considered behaviour, gave little or no attention to interdependence among behaving creatures.[1]

A new biology gradually established itself, during the nine or ten decades following White's death, but experimental work on behaviour, where it was undertaken, was external to this biology. (Douglas Spalding's studies of young animals were rediscovered only in the 1950s: 'I am inclined to think that students of animal psychology should endeavour to observe the unfolding of the powers of their subjects in as nearly as possible the ordinary circumstances of their lives', he had written.[2]) Significantly, and in Britain, attempts to resolve some of the problems associated with bird migration achieved comparatively little. The foundations of modern migration studies were being laid by Gatke, the Heligoland ornithologist, but Eagle Clarke, trying to follow in his footsteps in Britain, and most remarkably on Fair Isle, was denied the achievement his exertions might have brought him by a failure of imagination, as he thought of it, among British ornithologists. Those he was in touch with – and perhaps Clarke himself – underrated the importance of trained, synchronised observing.[3] Efforts were made to use questionnaires, with lighthouse keepers as the principal informants, by the Migration Committee of the British Association, but this showed up a defect in the Committee members and their advisers. They were without adequate techniques for digesting the data the questionnaires provided.[4] The eighteenth century saw the beginnings of epidemiology, but the first representative of what might be termed sophisticated statistical analysis in Britain, Francis Galton (1822–1911), was himself still discovering these techniques in the late nineteenth century.

In Britain, it was not until well into the first decade of the twentieth century that co-operative work on migration, including the vital ringing programs, began in earnest. We can say simply that ornithology remained too beholden to collecting and taxonomy, during most of the preceding era; or that the new biology – embryological, developmental and sooner or later evolutionary – had not brought with it a new and appropriate philosophy of biological science. The embryologists could isolate animals or plants as misleadingly as had the anatomists and classifiers. In the Preface to *The Botanical Atlas* (1883), Daniel McAlpine writes, 'Practical work requires to be *encouraged* for we have been so long accustomed to obtain our knowledge of Nature entirely through the agency of books, that the "book of Nature" has become to many the printed page and not the living reality.'[5] But what one did on a field trip with McAlpine was to gather more and better specimens. *The Botanical Atlas*, we find, is principally concerned with *micro-physiological* landscapes. The 'practical work' McAlpine was talking about was to take place, very largely, at the laboratory bench.

Even scientific history seldom progresses in a straight line. While attempting to engage life-processes, biologists commonly retained materialistic assumptions, during the century following that of Gilbert White; and as industrial changes were accepted, and a bourgeois respectability steadily extended its influence, the 'Design' conception – far from fading away – took on renewed vitality. By 1830, Sir Charles Lyell was arguing almost unanswerably for 'millions of years of prehistory', but a theory or notion of *successive* Creations – the work of the Deity following upon successive 'catastrophes' – could partially neutralise the challenge presented by these vast time periods. A Europe-wide romanticism seemed to threaten traditional religion, but Paley's *Evidences* and then the various 'Bridgewater Treatises', setting out the case for Design in fresh detail, were published during the first half of the century. Paley and the Treatises were consistent with a nineteenth century 'fundamentalist' Christianity. Even determined field naturalists could be confused, if not by the scientific materialism, then by a re-asserted biblical literalism. The extreme, but unavoidable, case is that of Philip Henry Gosse (1810–88), a highly talented observer who nevertheless remained a defender of fixed species until the end of his life. (In several respects, he was approximately the equivalent in Britain of the French entomologist, Jean-Henri Fabre, therefore.) With other fundamentalists, Gosse managed to avoid the facts the evolutionists were finding unavoidable, or else managed to fashion these facts into counter-evolutionary 'proofs'.[6]

Edmund Gosse tells us his father 'saw everything through a lens, nothing in the immensity of nature'.[7] Charles Darwin (1809–82) has never been accused of narrowness of vision (Gilbert White has, but quite misleadingly), but Darwin too suffered as a result of ambiguous conceptions of what science was or could do. *The Origin of Species* (1859) comes down on the side of materialism; but Darwin had what can almost be thought of as a second career, running parallel with the career for which he is most celebrated. As well as an evolutionist and morphologist, he was a behavioural and comparative psychologist. He produced the first methodical notes on the behaviour of very young children, and worked on ants, bees and earthworms, in this second capacity. In 1872 – in answer to the view that human emotional expression was without equivalents in non-humans – he published *The Expression of the Emotions in Man and Animals*, a study drawing largely on his own observations and notes. His 'two careers' were never satisfactorily united; until the closing years of the nineteenth century, 'behavioural science', the term, would hardly have had meaning; although, in this second career, Darwin too was a forerunner of the ethologists.

White foreshadows Darwin in the latter's lesser known role, and we can find points of similarity between the two of them also in the *Origin of Species,* the presentation Darwin himself could think of as an 'act of murder'. White did not foreshadow the natural selection theory – and indeed might have found that theory alarming – but he was already thinking critically about altered or altering adaptation. Guardedly, but in substance, he was already allowing for the instability of species.

2 Change and the limits of change

An example of what, to us, can look like evolutionist thinking appears in one of White's few comprehensive 'descriptions'; it was written in January 1789, and concerns the great northern diver. The bird had been caught alive on Woolmer Forest, and White, who had never previously met with the species, examined it thoroughly. Naming it as 'Colymbus glacialis Linn. the great speckled Diver, or Loon', and appreciating its dependence on under-water hunting, he goes on:

> Every part & proportion of this bird is so incomparably adapted to it's mode of life, that in no instance do we see the wisdom of God in the Creation to more advantage. The head is sharp, & smaller than the part of the neck adjoining, in order that it may pierce the water; the wings are placed forward & out of the centre of gravity, for a purpose which shall be noticed hereafter; the thighs quite at the podex, in order to facilitate diving; & the legs are flat, & as sharp backwards almost as the edge of a knife, that in striking they may easily cut the water; while the feet are palmated, & broad for swimming, yet so folded up when advanced forward to take a fresh stroke, as to be full as narrow as the shank.

The swimming of birds is a kind of 'walking in water', he says, but 'no one, as far as I am aware, has remarked that diving fowls, while under water, impell & row themselves forward by the motion of their wings, as well as by the impulse of their feet'. This can be ascertained if one watches ducks being hunted by dogs in a clear pond; and it brings him back to why the wings of diving fowls are placed 'so far forward'. This is 'doubtless not for the promotion of their speed in flying, since the position certainly impedes it', but is 'probably for the encrease of their motion under water by the use of four oars instead of two'. At the end of this description, predictably, he gives the measured dimensions and weight of the bird he has examined.[8]

We can find links of a sort between this and evolution theory. The nineteenth century evolutionists, Darwinist or otherwise, took elaborate structural adaptations as their data, and were originally directed to these adaptations by accounts such as this of the great northern diver.[9] But in White's own diary/notebook, what he is saying about the same bird represents a step back, more than a step forward. The 'wisdom of God in the Creation' again recalls John Ray, and in the passage just quoted White is assuming the peculiarly formed legs, feet, wings and so on of the great northern diver (the species) to have been devised at its 'creation' to fit it for under-water swimming. He could see divine workmanship, in the same way, in the equipment distinguishing tadpoles from older frogs. The 'fish-like' tails of tadpoles had been given them so that they could live in water, and by providential arrangement, when it was time for them to progress to a dry-land existence, their tails shrivelled and their legs grew.[10]

White was sixty eight when he described the great northern diver; but he could hold several perhaps conflicting views simultaneously. A received and reassuring Design assumption was among what I called the 'first thoughts' which could remain in his work alongside his 'second' or criticised thoughts. As a self-conscious and critical naturalist, he was much less sure of Design, or Design as all-explaining. Later in 1789, for instance, in September, he examined a landrail or corncrake. His description of this bird, though brief, is reminiscent of the description of the great northern diver, except that here the physical make-up of the bird is at odds with one considerable feature of its 'mode of life'. The corncrake is generally admitted to be a 'bird of passage', he writes, and 'yet from it's formation seems to be poorly qualified for migration: for it's wings are short, & placed so forward, & out of the centre of gravity, that it flies in a very heavy & embarrassed manner, with it's legs hanging down'.[11] Anatomy could never have become a major study for White, we said, because – however a physical examination may have worked out with the great northern diver – physical characteristics were often unreliable guides to life-styles. The diver was so adapted to its watery element that it 'could not stand or walk with any efficiency', but was only rarely found stranded on dry land; but the corncrake's migration was a feature of its usual proceedings. In *The Natural History of Selborne*, in the first 'letter to Barrington', the corncrake is listed by White as one of his usual summer birds.

The Design theorists assumed that a migratory bird would be especially well equipped physically for long-distance travelling. White's own fieldwork showed that this was not always true, and to find inconsistency in one creature, or one area of its behaviour and structure, was to allow that some of the areas where consistency did seem to obtain might themselves need reassessment. White found that 'behaviour inconsistent with structure' was far from uncommon. He too put off publication, but his private writings show no evidence of guilty forebodings. In cases like that of the corncrake, and as his second thoughts matured, he was 'hinting' at an established re-adaptation. In one respect, the perhaps inept corncrake compared especially well with the 'splendidly furnished' great northern diver.

White was aware – and could hardly have failed to be aware – of artificial selection. As a result of domestication, species 'commonly become bigger', and he apparently discussed selective cross-breeding with Mr Woods of Chilgrove.[12] He includes a concentrated, intelligent paper on domestic dogs, in the *Natural History of Selborne*, referring to the 'single or restricted origin' of the huge range of dog breeds with which we are now familiar. By chance, and because of his usual friendliness, he had been able to examine some 'Chinese dogs', brought into England by a young ship's officer, animals which had been bred over many generations to live and do well on a diet of rice meal. (They were bred for the table, in China.) In addition to their exceptional feeding habits, these dogs had 'sharp, upright fox-like ears', he says. Considering various bits of evidence, he thought it probable that domestic dogs of all kinds were descended from animals having much in common with the prick-eared sledge dogs of northern Europe and Asia. 'Hanging ears', he says, which are esteemed so graceful, are 'the effect of choice breeding and cultivation'.[13]

But physical alteration is given only occasional attention; and White was relatively uninterested in alteration which was 'artificial' in that it could have occurred or been brought about only in captive creatures. He concentrated on animals which had adjusted behaviourally to new or altered conditions, and in a number of the cases he refers to, the *altered conditions* had been artificially arranged; but even far-reaching behavioural alteration *could* occur in the wild, he became convinced. Here, the 'artificial' examples helped the naturalist examine a potential and natural flexibility, a flexibility which might come into play if environmental changes occurred.

We have seen that, literally or effectively, White re-located plants. He created miniature climates in his garden frames and struggled annually to change and improve his soil. Sensitivity to natural surroundings was a British ideal, but many in White's day could be simplistic, or extravagantly optimistic, in their attempts at re-locating. We are given an example in one of the *Selborne* papers. White seems to have kept away from menageries, but in 1768, while returning from a visit to Ringmer, he made a detour to visit the Goodwood moose, a beast imported from the forests of North America.

The moose was of special interest to Pennant, who wanted an accurate description of it for purposes of comparative anatomy (he was asking, for one thing, what it had in common with the extinct Irish elk); but the unfortunate animal had died by the time White arrived.[14] It was already beginning to smell badly – and it was surprisingly large – but he measured it as thoroughly as he could.[15] He had undertaken this visit to oblige a colleague and no doubt to satisfy his own curiosity, and in *Selborne* the paper on the moose encourages the general reader. But very little of White's book is altogether irrelevant to one or other of his 'theme topics', we said, and he made this examination too into an enquiry of his own. He considered the moose behaviourally and ecologically: he suggests possible reasons for its death, and for the languishing and decline of a companion of the same species, another female, which had been with it until the previous spring.

The moose had been expected to 'eat grass on the plain ground', in the Sussex park, he found. Because of the great length of its fore-legs and the shortness of its neck, it had been able to gather food only with great difficulty, therefore. As he learned from the Goodwood keeper, the animal had 'seemed to enjoy itself best in the extreme frost of the previous winter'. Its removal from the environment it was both fitted for and used to had been too sudden and too complete, White saw. What it was required to do in its new surroundings was what it was virtually unable to do, and what it could do, and was best at, had been of no use to it in these surroundings. It could swim well, for instance, and would browse on water plants in its original habitat, but it seems not to have had access to lakes or islands at Goodwood. A young stag, a red deer, had been turned in with the moose, we are also told, in the hope that there might have been a 'breed' between them; but their unequal heights alone must have made 'commerce of the amorous kind' impossible.[16]

But with more thoughtful treatment, the moose might have survived, White is implying. (In another context, he agrees with an 'ancient author' that in principle animals of any sort can be tamed.[17]) An animal or plant is always part of a larger complex, animate and inanimate, but this is not to say that one and only one larger complex will do. The vine coccus 'from Gibraltar' had survived at Selborne despite the relatively cold winters it had had to put up with there, and canaries, if born in the British countryside, might perhaps become hardened to our climate.[18] The Selborne 'bee boy' White describes, who tried to sound and even behave like a bee, had made himself so acceptable to bees and wasps that he could handle and play with them at any time in complete safety – unlike Gilbert White.[18] Selborne, for that matter, provided an example periodically of humans who seemed to have made a considerable adaptation in regard to climate. The gypsies who passed through the parish each year had come originally from 'warmer climates', but while other itinerants lodged in barns and cow-houses, 'these sturdy savages seem to pride themselves in braving the severities of winter, and in living *sub dio* the whole year round'.[19] Coming back to the non-humans, the goldfish and silver carp we can observe in Britain have many of them made a permanent adjustment to lower temperatures. Gold and silver fishes, White says, 'though originally natives of China and Japan, yet are become so well reconciled to our climate as to thrive and multiply very fast in our ponds and stews' – our outdoor, unheated, ponds and fish tanks.[20]

Talk of immediate or limitless adjustment, White treated as 'speculative' and irrelevant to field experience. Innate responses and needs delimited alteration, artificial or otherwise. (In Barrington's experiments with song-birds, he was overlooking the inborn needs of the nestlings: White would not have found it extraordinary that young birds reared 'without parents or instructors', though they made chirruping noises, developed no proper songs.) In much the same way, the resources of a given place or locality afforded only finite opportunities, whatever improvements had been made in that locality and however adaptable the creatures coming into it.

In his own parish, White could observe the effects of limiting processes on the sizes of plant and animal populations; apart from the seasonal fluctuations, the familiar densities of creatures were rarely exceeded. He knew he did not fully understand these processes. When the young birds left their nests at Selborne, bird populations could become temporarily enormous, even by Selborne standards; four hundred house martins might be associated with a single large farmhouse.[21] 'Jealousy' and territorial 'rivalry' encouraged the dispersal of a bird species, we saw, but these mechanisms alone could not explain the steadiness of population sizes in the long term. White returns to the alarming geometrical progressions which would obtain were no forces at work to counteract them in an exchange with Robert Marsham, in 1790. Four pairs of swallows build each year on his house, Marsham remarks, and 'If these produce two broods of five young, you see, Sir, one pair only, will in 7 years produce above half a million, 559,870 birds'; and yet the numbers which nest there every spring remain constant.[22] This discussion took place some years before the publication of Malthus's *Essay on the Principles of Population* (1798); and just such demographic calculations, for instance as they appear in the work of Malthus, were to influence the nineteenth century evolutionists.

Most remarkably, demographic calculations were to influence Darwin and A R Wallace, the other 'discoverer' of the natural selection process. For these enquiring naturalists, the gulfs of prehistoric time had opened up. Independently of each other, they asked not what brought about the deaths of the vast numbers of young birds, fishes or insects which did not survive each year, but what enabled the few which did survive to survive, despite the innumerable other contenders. White did not ask this question, or ask it directly, and when he wonders what determines each year which of the swifts, say, shall return to Selborne to nest, he can suggest no answer.[23] But he remained aware of population sizes and finite resources. He was not an evolutionist, but evolution theories would eventually find room for facts such as he was isolating. Even in his last years he was 'taking down remarks' on re-adjusted behaviour, and was still rejoicing over the survival of the smallest – or rather, the survival of small or weak creatures in situations in which, considering merely their size or physique, they might quite conceivably not have survived.

3 Migration and 'motives'

White sometimes seems to assume that morphological adaptations as they occur in the wild were originally provided more or less complete, where instincts and predispositions, though they too were 'created', were given originally in a rudimentary form. One of the most interesting of the many notes he made after the publication of *Selborne* (in January 1789), is to be found in the *Journal* for 1792. His newly fledged bantam chicks, he reports in the September of that year, have begun to roost on a high beam in the stable. In doing so, they have given evidence once again of the 'earnest & early propensity' of certain game birds – notably chickens, pheasants, guinea-fowl, turkeys and peacocks – to fly up into trees or to seek other high perches for the night. At some time and in some way, ground-nesting birds have had 'imprinted on their spirits' a dread of rats and other vermin, White thinks; but roosting 'on high' is only one of the ways in which they have acted on this aversion. Partridges settle close together in groups or coveys in wide open fields, well away from hedges, for their greater security at night. For this reason too ducks float in groups in the midst of large lakes or pools during the hours of darkness and danger, 'like ships riding at anchor'. These measures are now thoroughly consolidated in the species in question, he thinks, or are now inborn.[25]

In White's view, 'they who write on natural history cannot too frequently advert to instinct'.[26] The fears and expedients of the ground-nesting birds show us inborn habits and, it seems, one or more of the 'great motives' at work. So presumably does migratory behaviour, the performance White adverted to frequently throughout his adult life. For some theorists, migration as such brought the Design concept in question. William Derham had indicated some of the stresses underlying the hibernation-migration dispute. If migration represented an instinct, he wrote, it was 'a very odd instinct': it suggested that some birds required rescuing periodically – or that the Creator had nodded when fitting these birds for particular environmental conditions.[27] At his critical and 'scientific' level, White was a 'modern', by comparison with Derham. Starting from the hirundine controversy, but moving on to 'migration in general', he saw and said that birds may be well suited during one season, in a given locality, and quite ill-suited during another, and that birds of a given species may be *migratory* in one place or set of conditions and *non-migratory* in another. As we normally come across it, migration seems to represent a native tendency, none the less. Whatever the implications of this for the psychological animal or for the Design theory, the migratory need and habit is 'imprinted on the minds' of the birds in question.

A 'prince of personal observers', White was impressed by the effort seemingly expended by many migratory species, even if few of them crossed great oceans. The summer birds, at any rate, struggled to return to their usual haunts each spring. They failed to keep to their 'expected times and seasons' only if they were directly obstructed, and they sometimes arrived on time despite storms of rain or even snow. (One of the spring arrivals at Selborne, the chiff-chaff, could be relied on almost to the day, March 20–23.[28]) In October, when their second broods were fledged, the swallows and house martins would congregate and become increasingly restless, and would then depart en masse, with much the same 'ardour' and punctuality. They did so in 1774, a year when the gnats and flies continued to be out long after the hirundines had gone. Towards the end of that year, therefore, White remarks concerning the swallows and martins too that 'want of food alone' does not explain their departure.[29]

Taking all his records together, he found that the comings and goings of the migrant birds correlated with dates at least as reliably as with local temperatures or even food supplies. We can agree with this, and can now throw some light on the coincidence with dates. As instinctive, a migratory bird is self-propelled, but may well need various stimuli from outside to set it going. One trigger stimulus, it is now understood, is a specific period of daylight, a 'day' of a particular length.[30]

White did not know about the 'daylight' trigger – although he knew that a specific stimulus might elicit an array of conduct. But he recognised the 'ardour' of migratory birds, and recognised, again despite some misinformation, that depending on local conditions a particular species might be migratory or non-migratory.[31] Numbers of White's contemporaries viewed animals as thoroughly malleable, we said. Others, and no doubt the majority, still accepted an unaltering instinct such as we can ourselves gather from the writings of John Ray. Ray had spoken of instinct in animals as 'nothing but a kind of Fate upon them'; it was a fate affecting the creatures concerned but to which they could not contribute, and it kept members of a given species to one established pathway.[32] White was perhaps thinking of Ray when he used similar words himself, or ascribed them to 'philosophers'; but – at his scientific level – he, White, was departing from this received notion. Not only 'surprising' migratory travelling (such as that of the 'ill-equipped' corncrake), but migration in general, he is beginning to 'hint', might at some time have been an adopted or expedient measure. A bird might be quite unadventurous as long as it met with no obstacles to the fulfilment of its needs, but faced with shortages of food or nesting sites – or as often happened in reality, faced with the 'jealous' driving away process which meant that shortages were avoided – it might try other, different nesting sites or begin a more extended, even a 'more dangerous', travelling in search of food. Migration involves the whole animal, but White was both observational and comparative. The food motive, though fundamental to migration, appeared also in numerous other habits, and the migratory action and need itself, though it operated as one 'impulse', was traceable to both nativistic and environmental ('place and convenience') causes. Had environmental conditions not made a tactic of some sort necessary, we might never have heard of 'summer' or 'winter' birds.

In *Selborne* his remarks on 'instinct' and 'passion' appear more in the letters to Barrington than in those to zoologist Pennant. The miscellaneous brief items which virtually make up some of the letters to Pennant were what that gentleman was asking for, and White was more overawed by Pennant than by Barrington, titled though the latter was. The two sets of letters cover much the same time period, and by the late 1760s, when White began his exchanges with both men, he was already a naturalist of some experience. Apparently, he felt able to state his own views boldly and at greater length in the 'second' set of letters. Without the difference in his relationships with the other two naturalists, then, 'instinct' and 'passion' might have been prominent also in the first series of letters, those to Thomas Pennant.

White believed in propensity as well as automatic reaction; but there is still something mysterious about the 'two great' motives. In even the second part of *Selborne*, where 'food' and 'love' are given major roles, they make an oddly abrupt entrance; and in the twenty five and a half volumes of the *Naturalist's Journal*, although self-interest is referred to there, they are not named at all. The effects of food and love are that they 'support individuals and maintain species', White says. As much as any pre-arranged compatibility between a living thing and specific sets of conditions, these motives make for short- and long-term survival. Throwing doubt on, not a concerned Deity, but a Design by means of which all adaptation has been arranged, White is seriously qualifying John Ray, as he introduces them. And yet, 'food' and 'love' may have had one of their origins in Ray's published writings. Ray refers to something very like them in *The Wisdom of the God in the Creation* – although in a section of that work dealing not with animals but with human beings.[33] White found the great motive idea in Ray, I suggest, and re-applied it, in the *Selborne* papers. In the process of recognising change and adaptability – in the process of rejecting the 'pre-formed' and 'static' conception of animate nature – he reminds us that for him John Ray was still a stimulus and catalyst.

White's study of migration, above all, seemed to expose altered but heritable behaviour, and 'food' and 'love' allowed, or seemed to allow, him a notion of re-adaptation. He in fact alternates between two different ideas, I have said, as he considers the great motives; 'food' and 'love' as summarising not the *causes* but the *functions* of behaviour will be returned to in the next two sections. But 'motives' were frequently causes – dynamic and affective causes – for White. (Significantly, he can refer to 'food' and 'love' also as 'ruling passions'.)[34] In this connection, 'two' great motives were not sufficient, we saw earlier, and without being quite specific about it, he adds several others to them. There were responses which were difficult to reconcile with any number of motives in this sense; the instant crouching of a game bird or the reactions of the bereaved cat which was offered the leveret, whatever their origins, were apparently fixed and autonomous. He knew from experience that some members of a given species might be *less capable* of re-adjustment than others. (His 'intimate' observing ruled out species as an unvarying stereotype). And yet, versatility, and even an apparent choice-exercising, were observable in the field; some animals, anyway, could make their own 'rescue attempts'.

Versatility seemed to underlie migration, and migration brought one back to another important fact; that animals were rarely quite solitary. They were inter-communicating and inter-influencing beings. Like the human members of a parish such as White's own, 'animals' were affected by, but were also contributors to, the environmental conditions in which they lived and worked.

4 'Congenerous' but divergent species.

White could be misled concerning migration, but this should not disract us from what Walter Johnson calls his 'simple fearlessness'. To dwell on *divergent behaviour within species* was iconoclastic, given his eighteenth century and Anglican circumstances, but he directs us in addition to *closely related species which diverge from each other*. These are the 'congenerous' animals and plants we have touched on already, the related animals and plants which in important respects, and from birth, lead quite dissimilar lives.

White refers to these diverging species as anomalies, but he adds new examples to those he has noticed already throughout his later career. He refers to them first as he notes that of the various 'thrushes' to be seen in his parish, two species – the fieldfares and redwings – spend only the winters there.[35] His other 'congenerous' groups include the three or four hirundines, the three cricket species, the two sorts of deer, 'housed' and 'naked' slugs, bilberries and whortleberries, carrion and hooded crows and the spring and autumn crocuses. The members of these groups are in some sense family members, although – in White's experience – they do not naturally interbreed.[36]

The three cricket species are congenerous, but their habits and some of their inborn propensities in each case are suited to the diverse local environments in which they live. The largest of the three, the mole cricket, is also anatomically adapted to its particular habitat; its mole-like 'hands', notably, distinguish it from the other crickets. With several of the hirundine species, their usual ways of flying and their *wing forms* are suited to the speeds and altitudes they work at and the 'various insects' they hunt.[37] White is attracted to, but wary of, these congenerous species, but they may seem familiar to the later reader. Rightly or wrongly, they bring to mind the Galapagos finches, the finches which Charles Darwin examined in 1835 and which, with various other creatures, started him on the road to the natural selection theory. The Galapagos finches were closely related but were of perhaps thirteen different species and sub-species; their main distinguishing features were their greatly varying beak forms, thick or thin, 'hard' or 'soft', where the beak of a given species enabled it to deal with a particular staple food. The total range of the beak forms, and the distribution within the archipelago of various possible foods – flower seeds, nuts, insects and so on – were such that large numbers of the finches could survive in one isolated island group.[38]

White can be congratulated merely for having been aware of related species which varied in environmentally appropriate ways; he recorded his examples in a 'pre-evolutionary' age, and while never travelling far from his home district. But the varying structural adaptations within the 'families' he looked at were not what most gained his attention. The *behavioural* diversity of the animals and plants he was concerned with here frequently *far exceeded* their structural diversity – if structural diversity was evident. This was another view White sometimes felt unable to state quite openly;[39] but in 1767, in an early letter to Pennant, he had referred to the way harvest mice provide nests for their young ones as 'perhaps their best' specific distinction; and with the spring and autumn crocuses, 'sibling species' as they would now be called, their behavioural and ecological differences were, he believed, their only distinguishing marks.[40]

Darwin remarks in *The Origin of Species* (1859): 'he who believes that each being has been created as we now see it must occasionally have felt surprise when he has met with an animal having habits and structure not in agreement'.[41] Sooner or later, *comparative examination* undermines the Design concept. Darwin's evolutionary case, the case with which he replaced 'Design', was built mainly on anatomical evidence. The fossilised remains of huge antediluvian animals had unsettled him, but he wrote after examining the beaks of the Galapagos finches: 'Seeing this gradation and diversity of structure in one small, intimately related group of birds, one might really fancy that from an original paucity of birds in this archipelago, one species had been taken and modified for different ends.'[42] In his comments on 'congenerous' animals and vegetables White was suggesting that some related species which are now variously adapted may once have been if not the same then very close in all respects. The Selborne naturalist makes no explicit reference to new species, but he was here among a scattered, querying avant-garde. Common ground can be discovered between White and Charles Darwin, in this matter of diverging but related species; and because of the difference in emphasis between the two naturalists – one referring first to structure, the other concentrating on 'modes of life' – we can find common ground at this point too between White and various twentieth century evolution theorists.

5 Function, compromise and evolutionary change

White was analyzing the 'divergent congeners'. In any of the 'families', one of the better known members becomes the control against which the other members, or their behavioural adaptations, can be considered quizzically. If he only circles the idea of speciation (the emergence of new species), he implies that a species has a 'historical dimension', a history in the course of which it may have changed and in a sense progressed. The divergent 'family members' seemed to represent change – perhaps a form of 'dispersion'. As he considers them, indeed, these species bear a distinct resemblance to the various plants and animals he notices which, though quite *un*related genetically, have arrived at mutual compromises.

White was interested in 'utility'; and 'food' and 'love' have two different connotations in the second part of *Selborne*. They are sometimes propensities but they can also indicate the *functions* or *'purposes'* of behaviour. White could confuse propensities and functions;[43] but to know living things, he believed, we must be aware of not only what they do but also what their behaviour 'is for'. Awareness of functions went with the Design theory. It appears throughout White's own description of the great northern diver. (As Ray had insisted, nature or its Author 'did nothing in vain'.) At one of his levels, White challenges Design or all-explaining Design, but while doing so he continues his careful noting of functional and 'useful' attributes. Considered in this context, 'food' and 'love' do cover all habits or established habits. Directly or indirectly, all established habits *facilitate* or *have facilitated* the survival of individuals and/or the perpetuation of species.

White's animals and plants are not isolated specimens. Animals, most obviously, mate and rear young ones, and may be periodic or permanent members of a pack or flock or colony, but plants too need each other. White believes in the individual contributor, but individual*ism* appears only rarely in the living world as he describes it. (He refers to violent episodes, rather self-consciously, but it seems not to have occurred to him that the 'great motives' themselves might be bringing about the deaths of myriads of creatures every year.) Food supplies, partners and nesting sites are finite, and numbers of offspring may be very large, but behaving creatures show little evidence of a continuing 'war of all against all'.[44]

They commonly exhibit a mutual accommodating, on the contrary, where this is to all appearances useful. Some farm cats and farm dogs White knew of accommodated each other when corn stooks were being 'housed'; the cats caught only the little red mice and the dogs only the common mice.[45] In one form, instinct is 'blind to every circumstance that does not immediately respect self-preservation, or lead at once to the propagation or support of the species',[46] but the establishing and acceptance of 'cantons' by breeding birds minimised competition for the relevant resources. Cantons, or territories, kept birds of the same species apart. Similarly, if conversely, various quite 'incongruous' creatures enjoyed advantages as they subdued their mutual distrust and accepted a close physical proximity. The wagtails which picked flies from around the grazing cows were in most respects quite foreign to the cows, but 'interest makes strange friendships', White says.[47] He thought the chaffinches which gathered into flocks during the winter months gained indirectly from this winter flocking, we saw.[48] He could not explain the quarrelsome behaviour of the rooks in their nesting colony, but he believed he could throw light on the tendency of starlings to keep company with rooks, a flight of starlings often feeding with a flock of these much larger birds. Perhaps rooks, with their 'stout but sensitive' beaks, are better at locating food in the ground than starlings, and perhaps 'their associates attend them on the motive of interest, as greyhounds wait on the motions of their finders, and as lions are said to do on the yelping of jackals'.[49]

The compromising animals, while satisfying some of their desires, were resisting others, but the 'interests' of all parties were served by their behaviour. White does not himself refer to similarities between the 'compromising' species and the 'congenerous but divergent' species, but the two gave rise to similar results; both were alternatives to a chaotic free-for-all. In practice, both contributed to the work of 'natural economies'. Natural economies 'served purposes', they were simultaneously restricting and enabling; complex but articulated, they ensured that large numbers of animals and plants could maintain themselves and reproduce their kind each year in relative peace and security.

Interest – self-interest – throws light on the original acceptance of the 'compromise arrangements', White believes; these too can be viewed by him as clever stratagems; but as normally encountered 'outdoors', these too are examples of institutionalised – widely observable but often *un*considered – response. They were learned originally, but they now belong as much to populations and groups as to individual creatures.

As White also assumes, some at least of the compromise habits are now inborn. They are further examples of behaviour which is 'acquired' and inborn. He offers no opinion as to *how* the assimilation of acquired habits may take place, but he has no doubt that it can and does take place. Migratory behaviour may be local and yet (as we would put it) genetically inherited. In the course of his remarks on the 'Chinese dogs', he refers explicitly to effects of this kind. The willingness of these dogs to subsist on rice meal must have begun in some earlier generation as an acquired adaptation, he assumes, but when the puppies were first taken onto the ship which brought them to England they 'did not relish' meat or bones. At the beginning of the long voyage the puppies were barely weaned, he adds, and so could not have learned much about solid food from their parents.[50]

Functional criteria would be central to evolution theory, and considered in retrospect, the notion of *inborn but acquired characteristics* carries us into the evolutionary debates. The species we can now observe have arrived at their present multiplicity because of 'natural selection and chance variation', Darwin says, but to these causes he adds two others; 'sexual selection', performed principally by female animals as they responded to the display of the males, and what he sums up as 'use and practice'. Regarding the last – to which he gave increasing, not decreasing, attention as time went on – Darwin was close to Jean Baptiste de Lamarck (1744–1829), the French zoologist and 'transformist'. With the help of use and practice, Darwin thought, acquired physical or behavioural characters might so permeate the acquiring animal's biology as to be passed on directly to that animal's offspring.[51]

'Use and practice' is by no means a blind alley, but as is well known, the alleged direct transmission of acquired characters was tested in numerous controlled experiments during the early and middle years of the twentieth century (often with the help of the fast-breeding fruit fly, *Drosophila*), and was found wanting. There has been no worthwhile evidence to support this direct transmission. The explanation, the adduced causal mechanism, did not prove out, and as it was invalidated so too – though unscientifically – the behaviour supposed to have been due to this cause began to be ignored or passed over. For some decades, in the English-speaking world, to recognise the supposed effects was in itself to be tainted with the 'Lamarckian' error.

Yet, in all sorts of cases, it seems as if acquired and even innovative responses have produced changes in the genetic make-up of later generations. (The examples suggested by White are enough to show that the possibility should be taken seriously.) And in the second half of the twentieth century, the sting was taken out of this famous controversy. What went a long way towards rehabilitating the relevant effects, and broadening the notion of what can or cannot influence evolutionary progress, what went a long way towards showing that the kind of results we are referring to here can be accepted *without* an acceptance also of the Lamarckian mechanism, was the change in evolution theory brought about by the progress of *ecology* on the one hand and *ethology* on the other. Though acquired characters could not get into the species make-up by the route the Lamarckians believed in, they could make their impression and appear eventually in the species via environment, and most markedly, *socio-ecological* environment. If this roundabout route is valid Gilbert White is still, if not in the discussion, then pertinent to it. He describes 'sociality' and a consolidated reciprocal sharing, and as an outdoor observer, he is aware always of a general inter-relatedness.[52]

Nothing controversial is intended by the use of the term 'acquired' here. All behaviour has a genetic basis, but the 'genotype' (the individual's gene pattern) will include many possibilities not realised in the current behaving individual – and may include an ability to learn and adapt. Similarly, I am not trying to bring 'group selection' into what is in fact a straightforward claim. I *am* assuming that Darwin was correct in giving his main thesis two parts; evolution results from natural selection *and* chance ('random') variation. Of the two, natural selection is an effect of environment, and in particular, local environment. Evolutionary change presupposes changed genetic make-up; this is where 'chance variation' comes in; but environment tells us whether or not a new variation can go forward successfully. The survivors, as the evolutionist understands them, not only keep themselves alive but also leave descendants. ('Food' and 'love' are the vital considerations White saw them to be.) Surviving creatures must have a combination of talents, therefore; but an environment offers or fails to offer a creature opportunities to develop its talents. At a particular place and time, environment is a set of conditions, and perhaps an obstacle course, which the candidate must be able to deal with in maintaining itself and in reproducing.

The evolutionists were and are concerned with new species, and clearly, where new species have appeared, competition has been of crucial importance. Pressure of *rising numbers* is always, or is annually, an aspect of the constantly impinging environmental background. Nature's regulation of population sizes was among the puzzles White recognised and returned to but did not solve; 'dispersion' was only part of the answer, he saw; and though he sometimes recorded violent episodes, competition was a topic he tended to avoid. He did appreciate, though, that a living thing is frequently and perhaps invariably involved with other living things. By the 'fittest', Darwin himself often meant the 'best fitted for given surroundings', and evolutionists now accept that if competition is a reality, so are social learning and teaching, and so is the requirement to allow for – to fit in with – the lives of other creatures.

'Environment' is not a *fixed datum*, for the more recent evolutionists. It includes the physical, and much of the physiological, surroundings of the creatures concerned, and the 'communities' (of both animals and plants), populations and larger or smaller groups to which they belong. The unit which *is selected* is the individual, but selective pressure may be *exerted by* its 'socio-ecological' surroundings, where, however, the individual or what he does may marginally alter these surroundings. A social convention may support adjustments in an individual, let us say a newcomer, and its descendants, until genetic variation catches up with and consolidates those adjustments, but the convention itself will have grown out of particular behavioural responses. At a given time animate as well as inanimate environment may seem to be quite unchanging, but because of the interplay between particular creatures and their physical and cultural surroundings, even an acquired adaptation – becoming itself part of social convention, and being *reflected back*, as it were, on the creatures which originated it along with perhaps large numbers of other creatures – may contribute to evolutionary change. This is unlikely to appear merely from the study of fossilised remains, but as one twentieth century evolutionist puts it forthrightly: 'behaviour is a producer of evolutionary change as well as a resultant of it'.[53]

6 Ethology and ecology

Standing back from this revised Darwinism, we realise how considerable the revision has been.[54] Darwin conducted behavioural investigations, and he pauses over 'behaviour which is alien to structure', but he was unable to, or at least did not, make behaviour a prominent factor in his evolutionary scheme. His 'sexual selection' depended on behaviour or conduct, but it is touched on only very briefly. The instincts of animals, as well as their physical characters, are products of evolutionary descent, and Darwin devotes a chapter of the *Origin of Species* to 'instinct', but 'it is difficult to tell, and immaterial for us', he says, 'whether habits generally change first and structure afterwards; or whether slight modifications of structure lead to changed habits'.[55] Darwin shows little interest in the environmental, feed-back effect of a changed habit – and little awareness of the relative rapidity with which a changed habit can spread through a population. His notes on the communal behaviour of ants and bees are virtually his only discussions, in the *Origin*, of social membership and activity.[56] His 'materialism' was not the materialism of earlier centuries, we can agree; but – in contradistinction to Gilbert White – he describes comparative morphology as the 'very soul' of natural history.[57]

The examination of instinct was obstructed by a seeming contradiction in terms in the nineteenth century, one carefully-researched chapter of *The Origin of Species* notwithstanding. For those influenced by Paley and the Bridgewater authors instinct was a 'divine provision', and the later 'vitalist' accounts were hardly less obscure.[58] The alternative was a dependence on 'association', and by the turn of the century, processes of 'conditioning'. With the second, to explain instinct was, in large measure, to explain it away; for the associationist and reflexologist, 'instinct' was a mere product of learning – and had to be learnt over again by each generation. A place for instinct would be found within professional zoology, Darwin's own work on instinct would be appreciated, but only when (1) evolutionary selection itself was widely accepted, and (2) fieldwork on behaviour was allowed to be a valuable, even an indispensable, part of biological research.

These conditions began to be fulfilled during the 1920s and '30s. During this period, the early 'ethological' zoologists began elaborating their case – or cases, for modern ethology arose in several different countries. Rejecting mere conditioning, the ethologists might have rejected experimentation along with it. Their concern was with the behaviour of intact animals, whatever form this took and – in principle – wherever it was to be observed.

By the middle decades of the twentieth century, ethology was building on the area of intersection between field natural history, animal psychology, ecology and evolution theory. Tinbergen (1907–88) respected the field naturalist if only because 'the naturalist knows perhaps better than any other zoologist how immensely complex are the relationships between an animal and its environment, how numerous and how severe are the pressures the environment exerts'.[59] But the ethologists insisted also on 'big drives', innate urges directing and informing behaviour. Konrad Lorenz (1903–89), for instance, specified four big drives: love, hunger, fight and flight.[60]

Predisposition and function could be confused by the mid-twentieth century ethologists. Big drives have sometimes been ideological as much as behavioural constructs, and today, if big drives are not quite repudiated, they are viewed with distrust.[61] More restricted innate needs still feature in all ethological theory, but so does a range of ready-use, innate items of behaviour. While innate needs help shape animal life, so does the natural selection process. 'Food' and 'love' provide the 'purposes' apparent in the life-style of an animal or plant, but they do so in two capacities; they are appetites, where this implies a periodic need, and are also requisites for survival. As the second, and in a way, they too are among the 'numerous and severe' environmental pressures.

As the ethologists assumed, the involuntary 'imprinting' and 'releasing' processes had been gradually reinforced because of their survival value.[62] A living thing might be in *a state of transition* at a given time; because of the evolutionary process, closely related animals may diverge behaviourally, and animals which seem entirely remote from each other may have important habits in common.[63] In some instances it is easier to recognise an animal by its behaviour pattern than by its skin or coat, or the muscular and skeletal apparatus including claws, teeth, horns or beak, still loosely referred to as its 'structure'.

Ethology has itself evolved. If big drives have been demoted, the study of 'emotion' has risen in status in recent decades. At a given time, a mood – a prevailing affective tone or tempo – may manifest itself in an animal's performance. (Behaviour which is 'inconsistent with structure' may not be inconsistent with, to take one instance, the operation of the subject's endocrine glands.) As Julian Huxley (1887–1975) observed, 'It has somehow been an advantage, direct or indirect, to birds to acquire a greater capacity for affection, for jealousy, for joy, for fear, for curiosity'.[64]

We do not have to leave behaviour behind if we are to avoid atomistic conclusions. (If genes, so much discussed today, are atomistic, this by no means tells us that behaviour must be atomistic.) Ritualised performance, as this has been emphasised by all ethologists, is functional, and it has an emotional aspect. Huxley's paper on courtship ceremonial and consequent 'pair-bonding' in the great crested grebe is now viewed as seminal. Ritual and affect featured prominently in the reports of the primatologist and honorary ethologist, Jane Goodall (1934–). Goodall describes the excited, efficacious but largely symbolic charging and stamping by means of which dominant chimpanzees assert their authority in the group, and the equally symbolic stroking and grooming by means of which the other, 'lesser' chimpanzees advertise their submission to authority – and, as it were, control the dominant animals.[65]

Interaction of this kind was common in animal life, it seemed. In the study already mentioned, David Lack showed that a singing and posturing robin, though it might violently attack a stuffed robin placed in its territory, seldom needed to make physical contact with a living intruder.[66] The postures and calls of the successful territory holder were recognised badges of rank. The territorial adaptation, as well as helping the paired birds to support themselves and their young, was often vital to pair-formation; vigorous dispersion spread unpaired males across the countryside. Behaviour expressing relationships within groups and societies has been a feature of ethological and other related research. Lorenz's long-extended examination of his fluctuating but hierarchical colony of jackdaws, at Altenberg, became itself an influential symbol. More recently research has been conducted on 'social predators' such as spotted hyenas, which hunt in skilled gangs in the wild with a strict division of labour. Hyenas do this with such efficiency, it has been confirmed, that lions will follow hyenas to rob them of the prey they kill.[67]

A gang of hyenas is hierarchical and yet is a *quid pro quo* arrangement. Robbery apart, even the lowliest members of the gang – those furthest down in the 'pecking order' – receive some of the food they help to obtain and are given some protection. So-called 'network foraging', observable in various birds which search for food on the wing, must confer mutual benefits. Red kites and, more obviously, black-headed gulls search individually but while remaining within sight or sound of each other. In their cases, it seems, being able to converge rapidly when one of their number discovers food has advantages which outweigh the disadvantages.[68]

Protocols and symbols do not entirely rule out violent confrontations. In special circumstances, perhaps in connection with a valuable food source, wild chimpanzees will define and defend a group territory. Status systems within a chimpanzee group make for relatively peaceful co-existence, but as Goodall found to her horror, proceedings between groups of chimps may take a quite different turn; when aggressive display is met with aggressive display, a murderous blood-letting may follow.[69]

Communication, though, is crucial to even this 'rule-breaking' activity. The *diffusion* of information has been studied by ethologists, among others. In given circumstances, a member may move up or down within a population or cultural group, its status may alter; young chimpanzees have to be encouraged and corrected in the use of 'symbolic' responses. In jackdaws or jackdaw families, according to Lorenz, advanced flying techniques are both learnt and taught, and the ability to recognise 'jackdaw-killers' is passed on educationally. Female chimpanzees learn much of their parenting behaviour from older females; and chimps communicate with each other constantly, and communicate with, while also remaining suspicious of, baboons. Many birds respond to the alarm calls of other bird species, and some species have essential alarm calls in common.

Local dialects within the songs of birds are socially transmitted, they are traditions. As traditions they may be consolidated by means of the 'evolution' which occurs within the cultural environment itself, and – as part of the 'obstacle course' a surviving bird must be able to negotiate – may also have a selective effect on the birds concerned. (Animals contribute in various ways to the environments which also 'severely' impinge on them.) Peripheral individuals or groups are more likely to respond successfully to altered conditions than those at the centre of a population; but a habit produced artificially in a brown rat may soon appear in addition in that animal's associate rats. In the classic case documented by Fisher and Hinde, the trick of opening milk bottles on doorsteps, once it had been locally mastered, soon became widespread among English great tits and blue tits.[70]

Twentieth century ethology became experimental. Dealing with functional and 'multi-variate' conduct, and wishing to examine this conduct critically, the ethologists adopted comparative methods. Lorenz summed up ethology as 'the comparative study of behaviour'.[71] Tinbergen's demonstration of the releasing effect of the orange spot on the end of a herring gull's bill was comparative – and conclusive.[72] A gull lacking the orange spot, the alleged releaser, but complete in all other respects, did not elicit the pecking response from the chicks, he showed, and the most rudimentary wooden 'gull', but with the orange spot in place, produced the pecking response invariably.

Tinbergen never lost sight of adaptation; he was an experimentalist who was himself a field naturalist. In these activities, it has been said, he was 'one of the first to collect quantitative data, to do careful controls, and to influence others by his example that field experiments were valuable for dissecting both cause(s) and effects of behaviour patterns'.[73] In recent decades, ethology and '*behavioural ecology*' have over-lapped. The observable living thing (the 'phenotype') develops only as an environment is encountered, and to know an animal or plant fully we must know the total 'ecosystem' of which it is part.[74] The evidence provided by climatologists and soil scientists – regional, continental and even global evidence – may bear directly on the performance of a given creature.

The ecological emphasis brought with it sophisticated comparative techniques; among naturalists, ecologists were some of the first to make use of so-called 'statistical' or 'factor' analysis; and ecology has encouraged collaboration among researchers. The benefits of collaborative observing appear clearly from, for example, the work of the British Trust for Ornithology. The expansion and decline of populations are viewed today as important, even urgent, areas for research. Working throughout the British Isles, the BTO members contribute each year to ringing and census programs, and from time to time take part in investigations such as that set up in the 1960s to examine the connection, if any, between falling numbers of birds of prey, in various parts of the country, and the use of chlorine-based insecticides (such as DDT). As much as with the census or migration studies, enquiries of this kind can be managed only with the help of large numbers of scattered reporters.[75]

As understood in this book, 'science' aspires to quantitative precision, whether or not this entails the use of spring balances, test tubes or scalpels. Since the mid-twentieth century, the interpretation and analysis of field reports has been revolutionised. Micro-technology has made it easier to gather in and redistribute material from widely separated observers, and 'statistical' methods have furthered, and indeed allowed for, the constructive use of large quantities of data. The same methods have enabled investigators to get to grips with the *confused multiplicity*, as it can seem, of an animal's relationships. Even in areas where the chlorine-based pesticides were in use, birds of prey were exposed to *many potential influences in addition* to the pesticides. It had to be shown that the relevant chemicals were getting into the 'food chain' of the birds, and that this was a major cause of their decline. Conclusive answers were arrived at in this instance because, but only because, a statistical – in practice a quantitative, comparative and controlled – processing of large numbers of field reports could be carried out.[76]

7 White and an African grey parrot

Perhaps paradoxically, Darwin opened the way for twentieth century ethology and ecology. And if my summary is generally accurate, the ethologists and ecologists have been doing the sort of work, and sometimes precisely the work, Gilbert White hoped for and believed in. The mid- and late twentieth century naturalists may or may not have been aware of White; the extent of his influence is not the subject of this chapter, or this book; but while pursuing their own research – in their own groups and organisations – they have in effect re-run most of his enquiries and tried out most of his 'hints'.[77]

White found both coherence and adaptability in the 'life and conversation' he examined. Remarks on individuals can imperceptibly become remarks on species, and *vice versa*, in his writings; 'the house cricket', 'the missel thrush' or 'the cuckoo' are terms encouraging ambiguity. He can expect too much of 'adaptability'; but he also draws our attention to adaptability. In various connections, and repeatedly, he illustrates a continuity between 'animals' and human beings. An animal may in some degree consider and evaluate its surroundings, he believes. John Locke (1632–1704) had allowed non-human animals appetitive experience and a certain associational thought, but had also assumed a 'perfect distinction' to obtain between men and brutes,[78] and John Ray had viewed animals as 'sensitive' but non-intellectual.[79] The notion of animals as mere automatons – 'bare machines' – was never widely accepted in Britain, but in White's own day, Humphry Primatt, a campaigner against the maltreatment of animals, while he insisted on the suffering animals could experience, took it as axiomatic that they 'lacked faculties of language or reason'.[80] White refers to an apparent weighing up of options in both 'quadrupeds' and birds, and makes thoughtful, restricted claims for linguistic ability in animals. Some animal behaviour can conveniently be termed intelligent, he shows; it is the sort of behaviour we refer to as 'intelligent' when we find it in a human being.

An attitude sympathetic to *creative problem-solving* in animals became widespread during the second half of the twentieth century. During recent decades, this is to say, a swing of the pendulum has occurred within the discussion of 'animal intelligence'. Intelligence, like the use of language by animals, was irrelevant or absurd to the reflex theorists, and activity of this sort is controversial today in part because of the enthusiasm and *in*caution with which it can now be embraced. Some of the claims appealing to facts uncovered by Karl von Frisch, who studied communication among not birds or quadrupeds but bees, provide examples. Because of the peculiar dance performed in the hive by a honey bee which has discovered a food source, other honey bees can leave the hive and, perhaps without having previously visited the spot, find their way directly to this food source. The story – I make no attempt to do justice to it here – is an amazing one; J.L. Gould remarks that 'in terms of information capacity at least', the honey bee language is second only to a human language.[81] But this need not constitute an argument for thought, or conceptual communication, on the bees' part. Sophisticated though their behaviour can seem, a more likely explanation is that it evolved, by minute increments, as actions and reactions were picked out and reinforced by the natural selection process.[82]

Demonstrations of the ability of chimpanzees and other primates to learn and use elementary human languages, produced in the past three or four decades, might seem much more promising, but some of these have eventually appeared flawed. In part, this may illustrate merely the difficulty of contriving appropriate tests for inferential behaviour in animals; and work on or with Alex, the African grey parrot, has still to be mentioned. Birds have sometimes been thought of as poor relations where intellect is concerned, but Alex performs remarkably in 'double discrimination', matching and extrapolation tests. His investigator and usual companion, Irene Pepperberg, controls carefully against inadvertent cuing on the tester's part; but she believes that personal relationships with herself and her students over a number of years have contributed to the parrot's intellectual progress. Parrots respond to stimulating company. As Alex himself might have put it, 'studies that fail to take into account the ecology and ethology of the animal, either by design or default, are unlikely to provide knowledge of the complete extent of the animal's learning capabilities'.[83]

In White's time, as in that of Charles Darwin, anecdotes about thinking animals were common enough. In the 1760s the French forester-naturalist, Charles Georges Leroy, though he could also discuss animal habits perceptively, was writing of the law courts, marriages and parliamentary assemblies of birds and other animals.[84] Gilbert White distrusted anecdotes; but he was hard on both undisciplined credulity and a prejudging – *im*moderate – scepticism. He speaks of the 'clever and courageous' decoy behaviour by means of which a partridge draws a predator away from her chicks.[85] The comment is described by James Fisher, perhaps with some justice, as one of the earlier writer's rare anthropomorphic lapses. But White knows that the decoy response is universal among partridges. He knows that it also occurs immediately, or as soon as the young ones are approached, he has watched this behaviour in the field.[86] He lacked the modern time concept, but local history was among his concerns and interests. In referring to the decoy technique as 'clever', he means, I suggest, that insight at some time contributed to its coming about; the technique shows evidence of insight, irrespective of whether a given partridge is insightful in using it. What he is saying here is debatable, in this case, but not merely naive.

'Animal intelligence' does not require a methodology of its own. But when birds or other animals are carefully watched, White insists, 'intelligent' behaviour appears more often than the professional naturalists might expect. Incubating birds show little ability to count, he thinks, but even so 'abject' a creature as the tortoise 'possesses a much greater share of discernment than I was aware of'. On first reaching the top of the ha-ha wall, and without having been beyond it, the tortoise proceeded with great caution, as though appreciating the danger.[87] In special circumstances, an animal may take constructive measures at variance with its usual conduct: White cites the case of the flycatchers which sheltered their young from the sun's heat by hovering over the nest. Insight and 'passionate' readiness are often bound up together, for White. 'Invention', as distinct from mere awareness, perhaps occurs only in so far as an animal is especially aroused.[88] However this may be, an animal which is incapable of insight is vulnerable when faced with unusual problems. (As described by White, it is in a similar position to the animal which is highly developed physiologically but which has been removed from the habitat to which it is therefore suited.) And conversely, an animal which can exemplify insight, and which is faced with an obstacle, may take an innovatory – and perhaps indirect – route to a needed object or condition.

8 White and evolution

White does not think that 'insightful' animals understand the long-term effects of what they do, but plainly it is to an animal's advantage to be able to solve problems. 'Function' and 'survival', the notions if not the terms, are treated with particular respect in *Selborne* as eventually published, although, even in the eventual *Selborne*, both are topics which remain unresolved.

A successful animal or plant both maintains itself and reproduces its kind, White says. Irrespective of whether we agree with him on propensities, we can give him credit for this implied definition. He and the evolutionists of the past fifty years agree in viewing 'love', and most obviously the leaving of offspring, as essential to 'success', while the nineteenth-century theorists did not dwell on reproduction to this extent. Darwin believed that the choosing of sexual partners helped shape and direct a species, but he merely skirts the idea that, biologically speaking, survival without the leaving of offspring is a pseudo-achievement.

But the nineteenth-century 'background' included a general and pervasive notion of development; and Darwin had seen uncontaminated evidence of the millions of years of pre-history. He gave no room to an evolutionary drive; but recognising both environmental pressure and varying genetic legacy, he described a conditional, selected descent and proliferation. In the twentieth century, principally because of Darwin's synthesis, the fact of evolutionary change could be taken virtually as given. During the past half-century, in particular, the biologists and naturalists making constructive contributions to evolution theory have added a more complete knowledge of essential biological conditions – behavioural and ecological, as well as physiological and genetic – to evolution theory as the Darwinists left it.[89]

Behaviour could carry the modest, if also 'fearless', White quite beyond what he could understand, but he sensed that findings such as his own were important. He willingly recorded examples of 'locally adapted instinct'; in *Selborne*, he assumes the nest-building of the suburban chaffinches and wrens to be innate, although on his own evidence it could have been learnt by, for instance, first-year adults from older partners. His puzzlement concerning the regulation of population sizes was given a harder edge in his correspondence with Robert Marsham, but he had referred to this difficulty already in *Selborne* and, before this, in various *Journal* entries. At one of his 'levels' – and foreign though this may be to a traditional notion of his outlook – he was among those who were challenging Design or Design as a biological explanation. In his day there was no received evolution theory – and more to the point, no natural selection and chance variation theory – to which new facts and insights could be added, and even allowing for his work on 'instinct' and migratory birds, he offers no satisfactory alternative to Design. But he shows up the essential weakness in the Design principle; and his was a challenge appealing to 'life and conversation'. We find it in his accounts of what are or may be altered inborn responses.

Chapter 11

'LIVING MANNERS'

1 Benefits of restraint

White does not provide a satisfactory alternative to Design, and – unlike Darwin or Wallace – does not attempt to do so. Typically, and even allowing for the array of records and results we have looked at, he is as remarkable for his reticence as for what he has to say. He had faith in the future of his discipline, and at the beginning of the twenty-first century he can still surprise us. In a sense, we can include him in discussions which had yet to take place in his day. But in White the natural historian, enthusiasm goes hand in hand with self-restraint. He knew he could not 'travel in time'.

He suggested lines of enquiry that other naturalists, and perhaps future generations of naturalists, might pursue, but he had little bent for the constructing of comprehensive theories. Just as relevantly, for having no bent for it has not stopped lesser men from theory-building, he saw that he hadn't the evidence for all-resolving, comprehensive claims. He believed in thoroughness, and he compensated for his own biases as far as he could, but he did not prejudge what would be achieved only when teams of researchers were in the field. He regretted his own lack of reliable co-workers, but as year followed year, and as he sometimes had to wait not years but decades for evidence vital to a particular enquiry, his own working attitude remained steady; 'natural knowledge' should never outrun inference from tested observation.

His enquiries remained unfinished. Even when organised teams were at work, he thought, the stage would never be reached where there was no more work to be done. Regarding his detailed paper on the house martin,

> My remarks are the result of many years observation; and are, I trust, true in the whole: though I do not pretend to say that they are perfectly void of mistake, or that a more nice observer might not make many additions, since subjects of this kind are inexhaustible.[1]

This was not an expression of despair; without knowing just where enquiries such as his would lead, he continued examining and reporting until the last weeks of his life. The natural world was often puzzling, but the egg-laying behaviour of the cuckoo, for instance, reminded him with a frisson that nature could 'still astonish us in new lights, and in various and changeable appearances'.[2]

Advantages went with his restricted and economical procedures. We can return for a moment to his paper on the 'Chinese dogs', the dogs which had adapted as a race to a diet of rice meal. It may have seemed to White that the dogs which originally accepted the rice meal had transmitted this habit to their offspring directly. Such a direct passing on was contemplated by various biological writers by the end of the eighteenth century, and it had for years been popularly associated with the development of the larger or faster farm animals or race horses. But what Gilbert White reports concerning the Chinese dogs is what he has seen and what he knows to have taken place. It is now generally agreed that acquired characters cannot be directly transmitted, but White is not discredited along with this theory. In connection with the Chinese dogs, he is to be commended, rather, for the carefully factual account he leaves us.[3]

His unwillingness to offer speculation as evidence gave him his immunity, in this matter of the Chinese dogs. Typically, he did not guess at what would be revealed only by further observation and analysis, and did not fill out the gaps in his observational material with received or deduced data on non-observables. Where Ray had moved easily between observed creatures and the so-called vegetative, sensitive and rational 'souls' of plants, animals and human beings, White does not allude even to the 'animal spirits' accepted by many of his contemporaries.[4] He makes little attempt to relate habits to bodily organs, brain, heart, stomach, 'parts of generation' and so on, or to the neurological basis which was beginning to be exposed in his time. In his references to 'hearing' in insects he was not deterred by the *lack* of physical evidence. He would not be side-tracked for long into 'descriptions', and he offers no views of living things appealing to chemistry or mechanics. In his version of 'scientific parsimony', behaviour and qualified forecasts are the limiting considerations.

Whether what a animal or plant does is 'mental' or 'physical' is eclipsed in his work by whether reliable patterns can be discovered in what it does.[5]

He apparently made little or no use of the telescope, and he was wary of the microscope; the use of the microscope should be discouraged in young naturalists, he suggests in a private letter.[6] By the standards of present-day ethologists, he was unnecessarily restricted to behaviour. (Today, ethologists refer without embarrassment to the work of endocrinologists, say, or neuro-physiologists, as well as to their own observing and testing, and some do discuss mental states.)[7] The restriction kept him clear, however, of arguments such as those we touched on earlier concerning reproduction, and clear of the religious entanglements which were never far from 'natural history'. His first thoughts remained along with his later thoughts, in *Selborne*, and John Ray was a continuing encouragement, but White did not seriously address the question of 'nature as an allegory'. He flirted, but no more than flirted, with the *virtual pantheism* which was in vogue in Britain during his later years. This pantheism is allowed a voice in various of his poems. Concerning the spring and autumn crocuses, he could ask:

> Say, what impels, amidst surrounding snow
> Congeal'd, the *crocus'* flamy bud to glow?
> Say, what retards, amidst the summer's blaze,
> Th' *autumnal bulb*, till pale, declining days?
> The GOD of SEASONS; whose pervading power
> Controls the sun, or sheds the fleecy shower:
> He bids each flower his quickening word obey;
> Or to each lingering bloom enjoins delay.[8]

Here, the natural world as an arrangement of pre-formed parts has been replaced by the natural world as a divinely indwelt totality. The second as well as the first could encourage fatalism, or in White's time the notion that 'whatever is, is right'. And working in much the manner of a more recent 'eco-soul', an all-pervading God of Seasons easily explained the divergent life-styles of, for example, the spring and autumn crocuses. 'Explaining' of this kind may have satisfied uncritical or merely fashionable naturalists, but in White's case the poetry on which he rather prided himself was a different order of work from his investigating. I shall return to the point later in this chapter. Even if he had been tempted to appeal to a 'pervading power' as a naturalist, we can say, the behavioural and comparative criterion, operating as a kind of censor, would have obstructed such a move.

But behaviour, if it excused him from arguments he was wise to keep out of, was also a releasing subject matter. Persistently 'inquisitive', in one of his hats, White welcomed the 'various and changeable appearances' with which the living world could confront him. On the strength of his behavioural and comparative stance, he challenged and quite departed from the views of many of his contemporaries. In these chapters, I have been looking at questions concerning the *historical explanation* and the *scientific verification* of his work, where these can be thought of as effectively distinct. On the one hand, what are the sources to which his outlook as naturalist can be traced. On the other, have the results of his factual 'enquiries' been corroberated in the findings of other and perhaps more experienced reasearchers? The second was one of the main subjects of the previous chapter. In the course of their own work the 'ethologists' and ecologists, in particular, have replicated his enquiries. He comes through this verifying process pretty well; if I am right, indeed, the later naturalists alert us to aspects of what he did – including aspects of his methods – which have frequently been overlooked by commentators. But now the *historical, explanatory* questions again demand attention. The ethologists and ecologists are, or were, two hundred years away from Gilbert White; they are separated from him by vast social changes and by more than one 'biological revolution'. If there are important points of agreement between the Selborne naturalist and these much later workers, how can this agreement – how can this foreshadowing – have occurred? I suggested earlier that the occurrence of White's 'exceptional' zoology and botany is not as mysterious as it at first seems. In the next two sections I shall look further at the cultural and psychological origins of his work, including the all-important behavioural emphasis. In the last section of this chapter I shall return to the advantages which went with this emphasis.

Historically speaking, White's meetings with Pennant and Barrington, and their requests for information, brought about the realisation of his career as naturalist; that career, exceptional though it was, was already a possibility in the 1760s. A British preoccupation with nature study in various forms and his own talents and susceptibilities led him towards natural history. His gardening and 'estate management' played their parts, as did the inter-active but conditional operation of the parish itself; we can find an 'empiricist' strain in his methodology and personal attitude, and he was by no means alone in questioning an all-effecting Design. But a further influence can now be added to these. In an age in which sociability and intelligent self-regard were highly valued, *human* 'life and conversation' was already the subject of scientific study. Moving to and fro between the eighteenth and the twentieth centuries is no doubt a dangerous proceeding, but White himself was not in danger. An eighteenth century figure with hopes for the future, he did not try to transcend what he could properly know. What he did as a naturalist was remarkable, but being able to see into the future – as opposed to having faith in the future – was not a prerequisite of what he did.

2 The 'science of man'

White took off in his own way, but took off, very often, from starting points suggested by other people. He separated himself from collectors, and for him taxonomy and dissection were never more than sidelines, but he was part of a broad eighteenth century involvement with natural history.[9] He knew that a sentient and generative – ultimately perhaps an *un*predictable – living thing was stirring within the biology of his time; he could at least negotiate works published in French.[10] He would have been encouraged by Reaumur, the great entomological observer, but he must have been ambivalent before the massive output of a Buffon. Buffon could recognise the mutability of species, albeit with careful reservations (and he viewed 'love' as a 'sovereign power'), but he could also exemplify just the approach to natural history White was explicitly rejecting. In his *Histoire naturelle* (1749–88), the celebrated Frenchman was trying to embrace the natural history of the world, and was willing to accept anecdotes and untested beliefs.

The British Enlightenment was part of White's 'background' (he was both progressive and in a sense reactionary). His observational and comparative method was his own, and was a British 'empiricist' method. His behavioural commitment was his own; animal and plant behaviour would be given the status in scientific circles he thought it ought to have only in the course of the twentieth century; but the behaviour of human individuals and groups could be considered in a manner we would now term scientific even in, or before, White's time.

An established British psychology influenced his work, I said earlier. This psychology recognised two constituents in its – human – subjects, namely *sensory and social experience* and *inborn appetite and 'passion'*. 'Experience' was indispensable to the mind's development – or mis-development. Conceptions were formed as the sensory world was met with or else were learned from other people; there were 'no innate ideas'. These conceptions could take on their own permanence and authority. (During childhood an individual might be seriously affected by bad schoolmasters or, perhaps most dangerously, elderly nursemaids.) But 'passion' was in the end decisive; at any rate, only an instinctive urge could counteract an instinctive urge. 'Reason', mere intellect, though it might influence action, could not initiate action. In Pope's simile, reason can trim the sails, but the passions drive the ship through the water.[11]

Behavioural evidence had always been a feature of this psychology; we learn more about a man by observing him, it was believed, than by listening to what he tells us about himself. John Locke frequently made use of behaviour or conduct. The position stated by Hume in the 1730s – whether or not Hume himself adhered to it – was one White was later to echo. In the study of mankind, Hume writes, we should glean up our materials from 'a cautious observation of human life, and take them as they appear in the common course of the world, by men's behaviour in company, in affairs, and in their pleasures'. Where evidence of this kind is 'judiciously collected and compared', Hume goes on, a 'science of human nature' can be founded upon it.[12]

Organismic 'systems' appear in British thought even in the early years of the eighteenth century. Observationally considered, Hume says, an animal or plant is a 'system'; the several parts both 'refer to some general purpose and have a mutual dependence on and connection with each other'.[13] But as it developed during the early and middle years of the century in Britain, the 'science of man' itself, the traditional psychology in sophisticated form, affected Gilbert White, I am suggesting.

This science appeared in a wide range of written work. In the *Essay on Man* (1733), Alexander Pope was trying to 'map' human nature, and elsewhere he discovers the 'ruling passions' of various of his contemporaries by observational and comparative means.[14] In an age which also saw the first blossoming of the novel, human nature was approached via carefully described individuals and the encounters between them. (The 'science of man' had little or nothing to do with typologies, or so-called 'character-writing'.) In *Humphry Clinker* (1771), Tobias Smollett presents a mixed assortment of travelling companions 'without commenting on them', but recruits his readers as observers and, again, comparative analysts.[15] Henry Fielding was concerned with 'not men but manners', and in *The History of Tom Jones* (1749) he adds that the novelist needs experience of contrasting professions and classes and contrasting conditions of life.

> So necessary is this to the understanding the characters of men, that none are more ignorant of them than those learned pedants, whose lives have been entirely consumed in colleges, and among books: for however exquisitely human nature may have been described by writers, the true practical system can only be learnt in the world.[16]

Knowing men by these 'natural' means, Fielding says, we can forecast their actions, where forecasting is at least as useful as moralistic judging.

Observable events need not be 'behavioural' events, but these writers were interested in behaviour. It was human behaviour; 'The proper study of Mankind is Man', Pope had said in the famous *Essay*.[17] To get from human responses and relations to those of animals and plants required a sideways leap, in the eighteenth century. White had Ray's remarks on behavioural study to help him. Further, the human individual as understood by the Augustans was socially involved, by definition; a 'propensity to company and society is strong in all rational creatures', and if men made society, society also made men.[18] Contributors to the science of man were to examine human communities, as well as individuals, and they should give special attention to 'primitive' communities. The county naturalists can be cited again here. In his *Natural History of Carolina* (1731), Mark Catesby had included material on the customs – and buildings – of the Indian tribes; increasingly, foreign travellers were becoming ethnologists.[19] In his *Natural History of Cornwall* (1758), William Borlase describes the games, festivals and mummers' plays of the indigenous Cornish people, and attempts a grammatical analysis of the ancient Cornish language.

Field studies such as these were closer to outdoor work on animal behaviour; but a gap had still to be crossed. Gilbert White crossed the gap. He addressed himself to animal and plant behaviour as a career, albeit in early middle age. Implicitly, he offered a rationale for this career. Outdoor and analytical work on non-human 'modes of life' was needed, because the behaviour of animals and plants constantly inter-penetrated with that of humans, and because, more than this, zoology and botany were demonstrably incomplete without the study of these modes of life. Animal conduct, notably, was both more intricate and altogether more varied than the current experts supposed. Animals showed evidence of gregarious need and felt, 'passionate', experience.[20] An animal's lifestyle was neither altogether learnt nor altogether inherited genetically, and observable responses we would usually associate with not only flexibility but problem solving were to be found in many animal species. The outdoor observer would meet with a sometimes disconcerting living world; perhaps most remarkably, behaviour might contradict structure; but if the personal and collective lives of human beings could and should be observed, and more than this scientifically examined, the same was true of the personal and collective lives of even common – and small – quadrupeds, birds, plants and insects.

3 Lateral thinking and calculated testing

White avoided, or tried to avoid, the prejudgement of observable events. 'You are better able to see with your own eyes than any man I know', he was told by John Mulso.[21] 'Lateral' thinking, as I put it earlier, was essential to much of his work.[22] He questioned long-established farming methods, and though swallows had 'always' nested in the large brick chimneys of the Selborne cottages, he was puzzled by this behaviour. (The swallows accepted what must have been uncomfortable nesting sites, he suggests eventually, for much the same reasons that the 'wild' missel thrushes accepted nesting trees close to human dwellings.[23]) On one occasion after another, he 'saw' on his own account the relevance of an *atypical* response to the *usual* behaviour described in most of his reports. What we are now likely to think of as some of his most remarkable 'hints' were the result of his special sensitivity to events another observer might have noted merely as atypical, or might have passed over altogether.

He looked for evidence for or against particular hypotheses, but he walked and rode about his locality too in the spirit of the mid-twentieth century naturalist who protested, 'birds are not hypotheses'.[24] He would have approved of the journal *British Birds*, with which the name H.F. Witherby (and thus some of the first systematically organised ringing work in Britain) is still associated, and in which for virtually a century the reports of amateur and part-time bird-watchers have appeared alongside those of professionals. Behaviour met with by accident and carefully watched and recorded – such as fish-catching on the part of crows, which have occasionally been seen jumping into a river feet first, and then emerging and flying off each with a fish – *is* of importance, if only for this reason: it may be that a great many crows are capable of catching fish, and that ornithologists have not known about this because, simply, no one has thought to explore this possibility. White would have been pleased but not surprised to learn that we can now make sense, or some sense, of the 'quarrelsome' behaviour among nesting rooks he recorded. And he saw that often the genuinely exceptional, if carefully noted, could illuminate the usual. The hop plants which had been lacerated by the hail told him something about the other hop plants, those which had remained untouched. The withdrawal of the swifts in August, in warm or even hot weather, perhaps threw light on the mass withdrawal of the summer birds – even though this took place only later in the year and in what was usually much cooler weather. The hibernation theory was weakened, as the swifts were considered; we are back with one of White's recurrent topics here; but that he could *expect the unexpected* is one key to the progress he made within these recurrent topics.

He could make comparative judgements such as these immediately, and almost without thinking, we come to believe. But he also deployed the exceptional and sometimes artificially produced conduct he recorded, or made 'experimental' use of it. His experimentation, as I am calling it, has been obscured for perhaps the majority of his readers by the brilliance of his first-hand reporting, and until well over a century after his death even those other naturalists who valued behavioural data showed little or no interest in the comparative handling of it. Benjamin Stillingfleet had placed two sets of seasonal observations from two distinct regions and climates side by side, but so as to illustrate the same point twice (local observations 'could be seasonal guides' for farmers and gardeners). He was not looking for something neither set of data provided on its own. Stillingfleet may have been a Gilbert White catalyst, but White, original almost in spite of himself, compared observations so as to expose what would *not* otherwise have appeared. Considering maternal animals or migratory and non-migratory birds; exploring behavioural 'puzzles', and challenging the usual time and place relations; he asked how *different* creatures behaved in the *same* conditions, and how the *same* creatures behaved in *different* conditions. (Again, it sounds simple once it is expressed.) On occasion, merely that he juxtaposes quantitative records is significant. The list of insect-eating birds which were *residents* in his part of Hampshire, printed in the first 'letter to Barrington' beside the list of (mainly insect-eating) birds known to be summer migrants, itself shows us the intelligent, 'inquisitive' naturalist at work.

It is not the scientist's results that mark him out as a scientist but his methods, his ways of getting to his results. Some of the minor tests White calls 'experiments' would not now qualify for this title; shouting at the tortoise through a speaking trumpet, we saw, gave him little definite information about that animal's hearing.[25] In the field, he rarely intervened at all. We find no mention in his writings of dummy eggs manufactured by himself, or cardboard silhouettes of hawks in flight, although he more than once removed old house martins nests from under his eaves, so that he could watch these birds building up their nests from scratch and compare their laying dates with those of house martins using old nests. He largely lacked the co-researchers necessary to real progress in the projects he had embarked on. But he was already analysing behaviour, and doing this without departing from public evidence; and typically, in the course of analysing, he made use of 'controls'.

While testing, he could neutralise the circumstances not being tested. His *neutralising* or *controlling* has often been overlooked. It is perhaps unlikely that an eighteenth-century parish naturalist should have made use of this procedure. Moreover, White's inferential steps are so briefly indicated that in some cases his 'use of a control' can seem little more than a happy accident. But the 'happy accidents' have to be added to the various examples we have ourselves looked at; the bullfinches fed or not fed on hemp seed, the testing of the tin chimney in one of the hot-beds and forcing frames, the cuckoo with a protuberance in its belly or the suggested experiment with the two sponges. In these, the intentional use of a control, as we would call it, is unmistakable. That event A is commonly followed by event B need not mean that A is a significant cause of B. The use of a control tightens up the elimination process; indeed, it makes methodical elimination possible; and it too weakens the authority of mere conviction. As with publicity itself, it helps the scientist towards a peculiarly refined 'accuracy'. The other researchers of White's time, we said, with honourable exceptions among the representatives of 'medical arithmetic', made little use of this double provision; Stephen Hales included controls only rarely, and Watson, of the *Chemical Essays*, makes no room for them at all.[26] White says nothing directly about the controlling process, but it was perhaps his view that the investigator of 'life and conversation' had special needs here. Whether or not this was true, and at one of his 'levels', his use of this check to simplistic generalisation seems almost to have been second nature with him.

4 A sensitive science

'*The Natural History and Antiquities of Selborne, in the County of Southampton*' was published at last, in January 1789.[27] But as I am viewing the great book, it represented, rather than constituted, White's achievement. This achievement appeared in what he did as a scientific naturalist, over a period of some forty years, and in what he avoided doing. He investigated the lives of his subjects without detracting from those lives; he was a scientific naturalist but one showing no tendency to 'atomise' or 'mechanise' living things.

Dissertations were published on the 'duty of mercy and sin of cruelty' to non-human animals in the second half of the eighteenth century, but during the same years sporting entertainments were almost invariably 'streaked with blood'. The representative countryman's response to an unfamiliar bird or animal was to try to shoot it; when six of the rare black-winged stilts appeared at Frensham Ponds, in April 1779, the pond-keeper immediately shot five of them, although 'after he had satisfied his curiosity, he suffered the sixth to remain unmolested'.[28] White's own feeling, though seldom explicit in *Selborne*, is expressed plainly in a letter to Robert Marsham: 'there is too strong a propensity in human nature towards persecuting and destroying'.[29] After his college days, White gave up even the angling he had once enjoyed, and in stormy weather he worried about voyagers at sea (such as Banks and Captain Cook). When he watched Blanchard the balloonist sailing high in the air – although he was prepared for this event, and registered the details with his usual exactness – his heart 'beat with joy and fear at the same time'.[30] He identified with the wagtails which ran in and out 'under the very bellies' of the cows, or the young frogs as they marched up Northfield Hill.[31] He frequently watched his behaving subjects at very close quarters, we have said; moving about the parish, on foot or on horseback, he did not empty the fields and hedgerows of animal life as he approached. The shy grasshopper warbler is generally heard rather than seen, and in the experience even of most naturalists it skulks in undergrowth or bushes; but 'in a morning early, and when undisturbed', we are told by White, 'it sings on the top of a twig, gaping and shivering with it's wings'.[32]

He attributed intelligence to numerous birds and quadrupeds. Where the Design fundamentalists saw man as having been given 'dominion over' the other living things, he could treat industrious or sagacious animals, anyway, as partners rather than subordinates. In one mood, he perhaps thought of the *swallows, frogs and bantam sows* as delegated agents. He could produce cheerful parodies of himself and his Selborne relationships; his ironic sense is part of the total jigsaw. In the *Journal* – and the coda to the *Antiquities* – he can describe Timothy the tortoise in mock-heroic terms.[33] In the poem *The Invitation to Selborne*, he refers to the 'mountains' and 'forests' of the Selborne parish, and to the 'wild majesty' of this little known region. His poetry was a social accomplishment, as much as anything else; we can imagine him reciting, rather as someone else might have played, at one of the gatherings of relatives and friends.[34] (He was pantheistic as, but only as, a poet.) He could look comparatively at the Selborne downs and woods; in a letter to a travelling nephew, he remarks that now he, the nephew, has visited real mountains, the Selborne hills will seem mere molehills.[35] This too was only half serious; in discussions of White, a single suggestion is never enough. His downs would remain beautiful, he was sure, whatever massive uplands they were compared with. A friend and niece, Molly or Mary White, perhaps came closest to understanding this.[36] His feeling for his home locality was itself one sustaining cause of his research 'as naturalist and scientist'. The patient sensitivity usually apparent in his attitude to non-human animals can partly explain his findings, the results of his research, 'scientific' though this was.

We can be misled by his domestic activities, even if they also prepare us for his later work. When he had no glass for a box frame he improvised using oiled paper, and his large garden frames, which had been made to his own specifications, were collapsible; in principle, when they were not in use, they could be taken to pieces and stored out of the rain. He enjoyed gadgets and, as a gardener, he was sometimes fussy over inessential details; his regular 'weeding woman' and the faithful Thomas Hoar must both have been very long-suffering. A mere acquaintance might have expected him to be the sort of tinkering – and in an important sense pre-judging – experimentalist he depicts unintentionally as he recalls Stephen Hales (or the latter's relatively benign summer activities). But again, a one-facet view is not sufficient. His journals are particularly neat, but he it is who warns us against becoming mesmerised by lists. As a devotee of natural knowledge, and though he was 'experimental', he kept aloof not only from bird-catching and the collecting of dead insects but from anything which might have insinuated itself between the naturalist and behaviour in a habitat.[37]

A talent for *data-processing* gave him experimental opportunities. He is too sweeping in his dismissal of the naturalist who 'works in his own study'; though his materials came from outdoors, much of what he did with them, over several decades, was done during the long winter evenings. Only, he too was making phenomenological 'maps', and his maps too would be open to challenge. Only behavioural evidence was allowed in; he draws attention to the instinctive responses of new-born animals, but leaves the 'secret' character of instincts quite out of account; and on the same principle, no behavioural evidence – and no sorts of behaviour, 'automatic' or 'intelligent' – could be summarily ruled out.

If results and methods cannot, after all, be kept quite separate, they need not be quite confused. White's work is full of 'happy accidents'. His method, the method he accepted, though it depended on *discrete bits of data*, enabled him, in Pope's phrase, to '*catch the manners living as they rise*'.[38] He was penetrating but not 'reductive'; he analysed the cat which fostered the leveret, with perhaps surprising results, but far from reducing mother-love to something other than mother-love, his analysis shows us the special intensity of this passion. He points to behaviour which is out of step with structure, but nowhere suggests that behaviour and structure must ultimately be opposed. He records disparities and even disharmony in the living world but in doing so does not make the parts of that world into quite self-sufficient entities. Today, he would have agreed that a living creature 'is not decomposable into independent bits of genetic coding',[39] and he has little in common with the early twentieth century research workers who turned behaving animals into laboratory 'preparations'. A machine-like animal or plant enabled Stephen Hales to answer some of the questions he had set himself, but at important points we can contrast White with a Stephen Hales. White kept out of metaphysics – and as a biologist he had no unifying theory – but his observing and analysing brings out the *restless continuity* of living things or nature at large. Individual creatures and local economies (with their soils and weather, their animals, trees and water mills) are systems but not closed systems, he shows. Often, important relations can be highlighted within the continuum only by means of data processing; but an inborn migratory habit may have had – as White sometimes implies, will have had – environmental conditions as some of its causes, and even an urgent, compulsive activity such as migratory travelling may require environmental triggers to set it off.

Preoccupied with 'behaviour', he hardly refers to the atomistic and mechanical theorists, but *he does not destroy his own case merely in the process of setting it up*. His bits of data are discrete, but discrete in the way that readings enabling us to draw a smooth graph are discrete. A behaving creature, though more or less integrated, may show evidence of altered adaptation. White does not try to show definitively how this can have happened, the Socrates of Selborne, he asks more questions than he answers; but he refers to a far greater range of responses than would be found by the casual observer. He implies that exceptional behaviour may illuminate usual behaviour;

he describes animals, in particular, as potentially active contributors; but they are active, on any occasion, within larger behavioural and ecological networks. The animals, plants and habitats he is working with may, any of them, affect each other – directly or indirectly.

9. The Wakes as it is today, from the 'great meadow'
Photo: the author

10. Honey Lane, one of the 'rocky, hollow lanes'
Photo: the author

11. Selborne church and vicarage in White’s time
Source: S. H. Grimm, *The Natural History of Selborne* (1789)

THE NATURALIST's JOURNAL. 117

Year / Place / Soil	Therm^r	Barom^r	Wind.	Inches of Rain or Sn. Size of Hail ft.	Weather.	Trees first in leaf. —Fungi first appear.	Plants first in flower: Mosses vegetate.	Birds and Insects first appear, or disappear.	Observations with regard to fish, and other animals.	Miscellaneous Observations, and Memorandums.
Selborne. Sept.r 6. Sunday.	58.	29 6-10 3/4	S.W.		fog. sun. pleasant. showers. moonshine.	rain in the night. Mushrooms begin to come.				
7. Monday.	63 1/2	29 7-10 1/4	S.W. W. W.		fog. sun. louring.	Wheat abroad injured; some begins to grow. Many hops picked. Mr Thomas Mulso left us, & went to Winton. Some wheat housed, but cold & damp.				
8. Tuesday.	62 1/2	29 8-10	W. S.W.	/	louring. sun. pleasant.	Gathered many cucumbers. gale all night. Brisk gale all day. Wheat housed.				
9. Wednes.	60. 65.	29 8-10	S.W.		sun. pleasant.	Hops are not large. The fly-catchers, which abounded in my outlet, seem to have withdrawn themselves. Some grapes begin to turn colour. Men bind wheat.—				
10. Thurs.	66.	29 5-10	S.		grey. sun. hot.	Wheat housed, & pease.				
11. Friday.	63.	29 8-10 1/4	N. N.W.	41.	rain. gale. sun. cool.	Ophrys spiralis, ladies traces, in bloom in the long Lythe, & on the top of the short Lythe.				
12. Saturday.	58.	30.	N.W.		cold dew. sun. bright even:	clouds. Some wheat is out. W. Trimming has a large field not cut. Gentiana Amarella, autumnal gentian, or fell-wort, buds for bloom on the hill.				

12. Page from the *Naturalist's Journal* for 1789
Source: The British Library

120

A man brought me a land-rail, or daker-hen, a bird so rare in this district, that we seldom see more than one or two in a season, & those only in autumn. This is deemed a bird of passage by all the writers; yet from it's formation seems to be poorly qualifyed for migration; for it's wings are short, & placed so forward, & out of the center of gravity, that it flies in a very heavy & embarrassed manner, with it's legs hanging down; & can hardly be sprung a second time, as it runs very fast, & seems to depend more on the swiftness of it's feet, than on it's flying. When we came to draw it, we found the entrails so soft & tender, that in appearance they might have been dressed like the ropes of an wood-cock. The craw or crop was small & lank, containing a mucus; the gizzard thick & strong, & filled with small shell-snails, some whole & many ground to pieces thro' the attrition which is occasioned by the muscular force & motion of that intestine. We saw no gravels among the food: perhaps the snail-shells might perform the function of gravels or pebbles, & might grind one an other. Land-rails used to abound formerly, I remember, in the low wet bean-fields of Christian Malford in North Wilts; & in the meadows near Paradise-gardens at Oxford, where I have often heard them cry, Crex, Crex. The bird mentioned above weighed seven ounces & an half; was fat & tender, & in flavour like the flesh of a wood-cock. The liver was very large, & delicate.

13. Page added to the *Naturalist's Journal* for 1789
Source: The British Library

Chapter 12

THE PERENNIAL NATURALIST

1 Natural science as a 'culture'

To be able to make some assessment of Gilbert White as naturalist, we must have considered his work historically, and must also have asked whether his findings are corroborated in the work of other naturalists and scientists. As appears from both these lines of enquiry, the present-day student of White is in an interesting position. White's work has been thoroughly replicated and its possibilities have been fully explored only during the past five or six decades. Because of the influences which helped form his attitudes, and because of what behavioural and ecological scientists, in particular, have been doing during these recent decades, we can appreciate his attainments more readily today than could his nineteenth-century readers – with their received biological concepts.

The behavioural and ecological science of the past half-century can help us do justice to White as naturalist; it suggests that a serious reassessment of what he did is long overdue. This need not involve 'illegitimate modernizing'.[1] The question of whether scientific work from one era can properly be compared with scientific work from another has been bitterly debated during recent decades. Some of the bitterness has been unnecessary. Definitions are vital, as usual, and have sometimes been blurred: *which* bodies of scientific work, we have to ask, and *which* eras? If nineteenth century scientific researchers could be affected by an intrusive 'scientific materialism', their successors in many cases have agreed (1) that the scientist qua scientist can never be omniscient, and (2) that none the less, and in an important sense, he can be objective. Along with experiment, publicity has again been the essential factor. It is axiomatic that mathematical endeavour can and should be, if not co-operative, then open to colleagues; the mathematicians have only to agree terms and symbols. Can we collaborate and corroborate, similarly, in investigating the phenomenal or existential world? Yes, say the post-nineteenth century scientists, in so far as we are concerned not with what the phenomenal world 'is', but with relative change or stability in that world.[2]

Similarities can be found between this and the thinking about science of the British 'empiricists'. In important respects the empiricists were closer to their twentieth than to their nineteenth century counterparts; thinking about 'science' has by no means developed steadily. To take a broader perspective, however, public evidence has had a place in 'scientific' research throughout the past three and a half centuries. It has been held that 'Scientific knowledge, like language, is intrinsically the common property of a group or else nothing at all'.[3] During the last three and a half centuries, and whatever else may have been entailed, quantitative data – measurables and measurements – have been an accepted scientific currency. Within this period, scientists have been able to understand and inter-act with their contemporaries in spite of national and political boundaries, and with their predecessors – and even, in a sense, their successors – in spite of historical boundaries.[4]

To compare research conducted within this period with much earlier work again, would often be problematical. And prevailing interests have varied, during the 'three and a half centuries': we could refer to the sub-eras of Boyle and Newton, or Darwin, or Einstein or, at the turn of the twentieth century, the eco-biologists and cyberneticists. The sub-eras, though, have not been quite self-sufficient. Researchers are not coerced by observables or measurables; but in practice, the attainments of the dominating theorists have each represented a culmination as well as a new beginning. Once a change has occurred, heroic forerunners can often be discerned. Charles Darwin found that as an evolutionist he had had dozens of partial precursors. A given forerunner may or may not have been an influence, he may or may not have been a contributor to the 'cumulative effect'; but quite properly, though with a certain awkwardness, Darwin added a new introductory section to the *Origin of Species* in recognition of these earlier writers.[5]

Modern science (that of the past three and a half centuries) has aimed at prediction, in any of the sub-eras; it has been asked in any instance whether the 'relative change or stability' represented a fundamental and reliable tendency. During the past century, the same science has sometimes arrived only at *un*certainty (John Locke insisted that two things 'cannot be both the same and different', but some twentieth century research seems to prove him wrong); but during the past century too a science concerned exclusively with phenomenal events has for the first time been generally recognised.[6] If I have described White fairly, he was an early and distinctive precursor of our own behavioural zoologists and botanists; only during the last hundred years, and most notably the last fifty, has the sort of nature study he tried to promote been taken up nationally and internationally. Macro-biology, at any rate, is still predictive. To be able to work with White as naturalist, we do not need to know all that was 'going on in his mind', or all the shades and aspects of his motivation. Whatever his other motives, and whatever the the total range of cultural causes affecting him, he intended his results to be used, and tested, first of all in connection with predictive behavioural enquiries.

2 The would-be team member

If nothing else – and his humanity can surely still reach us – a common currency, a code, links us with Gilbert White. We can work with him, outdoors and in the study, much as we can play music from a score written in his time, or, if the relevant notation has been left us, re-run and complete a game of chess started and left unfinished by eighteenth century players.

As it turns out, White foreshadows whole areas of later research. This need not have happened, but with his beliefs regarding method, and concentrating as he did on plant and animal behaviour, he was likely to have touched on many of the questions later behavioural naturalists have pursued. His descriptions of courtship feeding, for instance, show us an exceptional observational ability, but given this ability, and occupied with birds above all, he was likely sooner or later to have watched courtship feeding at first hand. The *Naturalist's Journal* was both a complex set of field records and a model of work in progress. Something similar is true of *The Natural History of Selborne*; it not only provides factual information concerning 'life and conversation' but shows its readers the sort of enquiries an observational and comparative attitude makes possible. In *Selborne*, White knew, he had touched only the edges of what might be discovered, even within a single locality, but if the twentieth century naturalists can help us to appreciate him, he from time to time throws light on them. His comparative aspect has often been overlooked, but bringing science to living things, rather than living things to science, he combined field observation with experimental enquiry. The same combination is now so widely accepted that it is in danger of being again under-valued; but as I suggested earlier, its recognition in the twentieth century came about only by stages.

Organised behavioural research had its beginnings in the final decades of the nineteenth century, but until the middle years of the twentieth, even leading ethological workers could seem unsure of their aims. Lorenz spoke scathingly of experimentalists, but he made frequent use of 'natural experiments'; while remaining an opponent of scient*ism* in ethology, he allowed friends and assistants to do quantitative testing for him.[7] On the other wing of the movement, Tinbergen implied at one point that only the physical, or physiological, could be objectively examined,[8] but he went on to become one of the great experimental investigators of behaviour in the field. (Tinbergen devised and used many pieces of experimental equipment, but he used them in investigations of conduct he had found to be naturally occurring.)

White foreshadows the practical attitudes of naturalists such as these. He tended not to intervene literally in the lives of his subjects, but he would have been glad to join forces with – or merely to work under – a Tinbergen. If he largely separated himself from physiologists, he was aware always of habitats as well as habits. He responded willingly to the special mobility of many bird species. He looked forward to the emergence of a widespread community of outdoor observers; he was or wanted to be the opposite of a 'hermit scientist'. He believed he should concentrate on one locality, as a student of behaviour, but – whether or not he enjoyed letter-writing – he viewed the regular writing and receipt of letters as of almost equal importance.

3 The remaining 'puzzle'

White was the original, and is the archetypal, student of the behaviour of intact animals. Some ethologists discuss mental states in animals today, but where the same workers are experimental – and where they offer their results to other experimentalists – they still depend on observable and measurable responses.[9] Some now avoid not only 'instinct' but 'appetite' as a term, but they agree that an animal will periodically be, in the Lockean phrase, 'uneasy in the want of' certain sorts of objects and experiences. As the field naturalists continue to report, behaving animals evidence fixed reactions but also inborn felt orientations and needs.

Today behavioural research is conducted by teams, which interact almost globally; county societies, TV companies, university departments and government ministries may all concern themselves with behavioural natural history. Gilbert White – despite what he says about 'true naturalists' – was often operating in fields of study which had still to be recognised as fields of study even by an interested minority. He was suited by behaviour. Behaviour, conduct, gave scope to both his sympathetic and his experimental needs; the man who sided with 'life and conversation', rather than anatomy and collecting, and the quantitative and questioning man could co-exist in this behavioural region. It is a mistake to try to smooth out White's 'mixed, uneven' character; to homogenise him is as bad as to recognise only a part of him, or to exaggerate a part and make it into the whole. But the man was sympathetic and objective in roughly equal parts; and – as it is now easier to appreciate – the *naturalist* was especially *coherent*. The naturalist's results are precise and often quantitative but in his studies of life and conversation the 'life' is not inevitably lost along the way. James Fisher remarks that, 'There are a great many naturalists . . . whose contributions to science stand higher than White's',[10] but if White's advocacy of *the observational and comparative study of behavioural adaptation in animals and plants* was not an important contribution, this was because most of his nineteenth-century successors, though they had easy access to the *Selborne* papers at least, were unprepared for this advance. *The Natural History of Selborne* sold in repeated editions between 1850 and 1900, above all, but its popularity during this period was due in part to its being thought of as *non*-scientific.

That of White was not predominantly a 'brass instruments' science, and by standards other than local standards, the Selborne winters were not cold, the hills of the parish were by no means high. But parochialism – or humourless, unbalanced parochialism – is absent even from his journal-notebooks.[11] While seeming to be unaware of them, he avoids the pitfalls he 'might have got into'; again and again, he sails with impunity between the menacing rocks or through the deceptive shallows. This has been a wide-ranging book, and various questions started in the course of it have not been followed up. In these last chapters, the question of how his achievement could have happened has been faced but only partially answered. He did not know where enquiries such as his would lead, but every paper in the *Selborne* collection was minutely considered and revised. As Walter Johnson says, he was 'never a victim of introspection', but even while trying to disguise the originality of his work, he remained convinced of its value. A selection of influences stimulated and combined in his 'scientific' outlook. Stephen Hales took his place with John Ray, the improving landlords, the Lockean thinkers and the observational students of human nature. White's *taste* stood him in good stead. He wobbled on occasion but always came back to the 'outdoor' examination we think of inevitably as his authentic occupation; the right influences combined in his attitude as naturalist.

I referred earlier to his ironic touches. Irony is present in his descriptions of the new, larger establishments set up first by Thomas White and then by Benjamin as they retired to become country gentlemen; he pointedly refers to Thomas by his newly adopted surname, Holt-White, in the final volumes of the *Journal*. But envy seems to have got no grip on the small man. As a natural historian, he could be goaded into defending himself; in TP xxxix particularly, he provides some indication of the assistance he has given Thomas Pennant. But self-promotion was not one of his various talents or interests; he tried to promote a natural history method, rather, and a general, provisional aim. He says towards the end of his paper on earthworms: 'These hints we think proper to throw out in order to set the inquisitive and discerning to work. A good monography on worms would afford much entertainment and information at the same time, and would open up a large and new field in natural history'.[12]

The quietly permeating effect of his contribution has no doubt been immense; but perhaps because of his willingness to be a mere team member, he has been under-praised or quite unacknowledged, if not by his successors, then by the principal figures among his successors. The *Formation of Vegetable Mould Through the Action of Worms* adds to the stature of Charles Darwin, and will do so however his evolution theory is modified, and the young Darwin read Gilbert White with pleasure, we know; but in this important ecological essay, White is noticed only in a brief aside and only dismissively.[13]

In these chapters, and while stressing White's 'modernity', I have perhaps been guilty of a comparable error. I have discussed him almost as if he were a twentieth century figure. But I have not meant by this that he can be annexed to the twentieth century. He cannot be separated from 'Selbornian scenes', but he will remain in the avant-garde. His fellow observers are catching up with him today, but they are unlikely to overtake him. The tortoise, we remember, is one of his band of insignificant creatures.

APPENDIX

Some relevant papers and extracts.

1 Dissection of a Live Dog

Stephen Hales, 'Vegetable Staticks', *Statical Essays*, Vol. I (1733), Ch.vi, experiment cxiv.

I tyed a middle sized Dog down alive on a table, and having layed bare his windpipe, I cut it asunder just below the *Larynx*, and fixed fast to it the small end of a common fosset; the other end of the fosset had a large bladder tyed to it, which contained 162 cubick inches; and to the other end of the bladder was tyed the great end of another fosset, whose orifice was covered with a valve, which opened inward, so as to admit any air that was blown into the bladder, but none could return that way; yet for further security, that passage was also stopped with a spiggot.

As soon as the first fosset was tyed fast to the windpipe, the bladder was blown full of air thro' the other fosset; when the Dog had breathed the air in the bladder to and fro for a minute or two, he then breathed very fast, and shewed great uneasiness, as being almost suffocated.

Then with my hand I pressed the bladder hard, so as to drive the air into his lungs with some force; and thereby make his *Abdomen* rise by the pressure of the *Diaphragm*, as in natural breathings: Then taking alternately my hand off the bladder, the lungs with the *Abdomen* subsided; I continued in this manner, to make the Dog breathe for an hour; during which time I was obliged to blow fresh air into the bladder every five minutes, three parts in four of that air being either absorbed by the vapours of the lungs, or escaping thro' the ligatures, upon my pressing hard on the bladder.

During this hour, the Dog was frequently near expiring whenever I pressed the air but weakly into his lungs; as I found by his pulse, which was very plain to be felt in the great crural artery near the groin, which place an assistant held his finger on most part of the time; but the languid pulse was quickly accelerated, so as to beat fast, soon after

I dilated the lungs much, by pressing hard upon the bladder, especially when the motion of the lungs was promoted by pressing alternately the *Abdomen* and the bladder, whereby both the contraction and dilation of the lungs was increased.

And I could by this means rouse the languid pulse whenever I pleased, not only at the end of every 5 minutes, when more air was blown into the bladder from a man's lungs, but also towards the end of the 5 minutes, when the air was fullest of fumes.

At the end of the hour, I intended to try whether I could by the same means have kept the Dog alive some time longer, when the bladder was filled with the fumes of burning *Brimstone*: but being obliged to cease for a little time from pressing the air into his lungs, while matters were preparing for this additional Experiment, in the mean time the Dog dyed, which might otherwise have lived longer, if I had continued to force the air into his lungs . . .

2 Bird Migration as a Fact

Mark Catesby, 'The Retreat of Birds of Passage', *Gentleman's Magazine* (1748), vol. xviii, pp 447-448.

The various conjectures concerning the places to which *birds of passage* retire, are occasion'd for want of ocular testimony to bring the matter to some certainty. The reports of their lying torpid in caverns and hollow trees are ill attested, and absurd; as is a late-broach'd hypothesis, which sends them above our atmosphere for a passage to their retreat. I cannot but agree to the general opinion of their passing to other countries by the natural way of flying, with this additional conjecture, that the places to which they retire lie probably in the same latitude, in the southern hemisphere, with the places from which they depart, where the seasons reverting, they may enjoy the like temperature of air. . . . The manner of their journeying to the southern abode may vary, as the different structure of their bodies enables them to support themselves in the air. Birds with short wings, as the Red start, Black cap, &c. tho' incapable of long and swift flights, may pass by gradual and slower movements; and there seems no necessity for a precipitous passage, because every day affords an increase of warmth, and a continuance of food . . . It is probable these itinerant birds may perform their journey in the night time, to avoid ravenous birds, and other dangers which day light exposes them to, which I have reason to believe from the following instance: Lying on the deck of a sloop, on the north side of *Cuba*, I, and the company with me, heard three nights successively flights of Rice-birds, passing over our heads northerly, in their direct way from *Cuba*, and the southern continent of *America*, from whence they go annually to *Carolina*, at the time rice begins to open, and after growing fat with it return south again. Thus our summer birds, when, by the approach of winter, they find a want of food, resort to some other parts of the globe, where they find a fresh supply.

The flight of *birds of passage* over the seas, has, by some, been considered as a circumstance equally wonderful with other stories concerning them; and especially in regard to those with short wings, among which Quails seem, by their structure, little adapted for long flights; nor are they ever seen to continue on the wing for any length of time; and yet their ability for such flights cannot be doubted, from the testimony of many. *Bellonius* in particular reports, that he saw them in great flights passing over and re-passing the *Mediterranean* sea, at the seasons and times they visit and retire from us.

As for winter *birds of passage*, these are but few, there being but four that I know of, *viz*. the Fieldfare, Redwing, Woodcock and Snipe, which two last I have frequently known to continue the summer here, and breed; so that the Fieldfare and Redwing seem to be the only *birds of passage* that leave us at the approach of summer, and retire to the northern parts of the y[e] continent, there breeding and remaining during y[e] summer, and at the return of winter are driven southerly in search of food, which the ice and snow, in those frigid regions, deprives them of. There are many others, particularly of the Duck and wading kind, that breed, and make their summer abode in desolate fenny parts of our island. When the severity of winter deprives them of their liquid sustenance, necessity obliges them to retire towards the sea in numerous flights, where in open brackish waters they find relief, and at approach of the spring return to their summer recesses.

The retirement of *winter birds of passage* is known to be *Sweden*, and other countries in that latitude; but as they would find them too cold and destitute of provisions, they journey gradually through the more moderate countries of *Germany* and *Poland*, and arrive not at these northern regions, adapted by providence for their summer abode, and breeding of their young, till the severity of the cold is abated;

when they revisit us in winter, they return back in the same manner.

The coming of these birds is then pretty well accounted for, but the cause of their departure is yet a secret in nature. In short, all we know of the matter ends in this observation: That Providence has created a great variety of birds and other animals, with constitutions and inclinations adapted to the different degrees of heat and cold, in the several climates of the world, whereby no country is destitute of inhabitants, and has given them appetites for the productions of those countries, whose temperature is suited to their nature, as well as knowledge and abilities to seek and find them out. From which we may infer that the birds we have mention'd could no more subsist in the sultry climes of the *Molucca* isles, than birds of Paradise in the frigid regions of *Sweden* or *Lapland.*

Besides the *migratory* birds, already mentioned, which breed and remain the whole summer, there are other birds that arrive *periodically* at certain places, for the sake of grain, and after no long continuance depart, and are no more seen till that time twelvemonth, as is observ'd of the Rice-bird, and Blue-wing of *Carolina.*

3 Reports of 'Torpid' Swallows

Daines Barrington, 'On the Torpidity of the Swallow Tribe when they Disappear', in *Miscellanies* (1781). Reproduced here without footnotes.

. . . I shall begin with the *Swallow*, as Mr.Pennant does in his British Zoology; and premise that I mean the species whose tail is most fork'd, and which is mark'd with a red spot on the forehead and chin.

This bird appears the first of its tribe, and (as I conceive at least) hides itself under water during the winter, because, in the few instances where the relator hath been able to particularize the species thus found, it hath happened to be a swallow.

There is scarcely a treatise on ornithology, written in the Northern parts of Europe, which does not allude to the submersion of swallows during the winter, as a fact almost as well known as their peopling the air during the summer; and because the name of Linnaeus is respected by most of the incredulous on this head, I copy from him the following words in the description of the bird. "Hirundo [*Rustica*], habitat in Europae domibus intra tectum, unaque cum *urbica* demergitur, vereque emergit."

It is clear from the expression of *demergitur* (though perhaps not classical) that this naturalist conceived these birds hid themselves under water during the winter; and it is to be observed that he seems to have stated it after a proper examination, because in the Fauna Suecia, published five years before, he omits the mention of this circumstance.

As the instances of finding swallows under water are most common in the Northern parts of Europe, I shall begin with the testimony of the inhabitants of that part of the globe.

Mr. Peter Brown, a Norwegian and ingenious painter, informs me, that from the age of 6 to 17, whilst he was at school near Sheen, he with his companions hath constantly found swallows in numbers torpid under the ice, which covered bogs, and that they have often revived upon being brought into a warm room.

Baron Rudbeck, a Swedish gentleman, who was not long since in England, hath assured me that this fact was so well known in Sweden as to leave no doubt with any one.

Mr. Stephens, A.S.S. informs me, that when he was 14 years of age, a pond of his father's (who was vicar of Shrivenham in Berkshire) was cleared during the month of February, that he picked up himself a cluster of three or four swallows (or martins) which were caked together in the mud, that the birds were carried into the kitchen, on which they soon afterwards flew about the room, in the presence of his father, mother, and others, particularly the Rev. Dr. Pye . . .

The compilers of the Encyclopedie (art. Mort.) have inserted the following observation and fact in relation to swallows discovered in the same situation. "Plusieurs oiseaux passent aussi tout l'hyver sous les eaux, telles sont les *hirondelles*, qui loin d'aller suivant *l'erreur populaire fort accreditée*, dans les climats plus chauds, se precipitent au fond de la mer, des lacs, & des rivieres, &c."

It is there also stated, that Mr. Falconet, a physician, living at Paris, had seen in one of the provinces, "une masse de terre que les pecheurs avoient tiree de l'eaue; apres avoir lavee & debronillee, il appercut que ce n'etoit autre chose qu'un amas d'hirondelles," which, on being brought to the fire, revived, the fishers declaring that this was not uncommon.

The late ingenious Mr. Stillingfleet informs us, that one swallow's being found at the bottom of a pond in winter, and brought to life by warmth, was attested to him by a gentleman of character.

Some years ago the moat of Aix-la-Chapelle was cleaned during the month of October, and the water let out for that purpose, when on the sides of the moat, and much below the parts which had been covered with water, a great number of swallows were seen to all appearance dead, but their plumage not impaired.

Du Tertre mentions, that a Russian of credit had told him, that, a piece of ice in a village of Muscovy having been brought into a house with swallows in it, they all revived.

There are several reasons why swallows should not be frequently thus found; ponds are seldom cleaned in the winter, as it is such cold work for the labourers, and the same instinct which prompts the bird thus to conceal itself, instructs it to choose such a place of security, that common accidents will not discover it.

But the strongest reason for such accounts not being more numerous, is, that facts of this sort are so little attended to; for though I was born within half a mile of the pond near Shrivenham, and have always had much curiosity with regard to the natural history of animals, yet I never heard a syllable about this very material and interesting intelligence till very lately.

To these instances I must also add, that swallows may be constantly taken in the month of October, during the dark nights, whilst they sit on the willows in the Thames; and that one may almost instantaneously fill a large sack with them, because at this time they will not stir from the twigs, when you lay your hands upon them. This looks very much like their beginning to be torpid before they hide themselves under the water.

A man near Brentford says, that he hath caught them in this state in the eyt opposite to that town, even so late as November.

I shall conclude the proofs on this first head by the dignified testimony of Sigismond King of Poland, who affirmed, on his oath, to Cardinal Commendon, that he had frequently seen swallows which were found at the bottom of Lakes.

I shall now proceed to the second species of the swallow-tribe, called a *martin*, which hath no colours but black and white, hath a shorter tail than the preceding, and builds commonly under the eaves of houses.

I may be mistaken, but I shall here again hazard a conjecture that this species does not hide itself under water during the winter, but rather in the crevices of ricks or other proper lurking places above ground, as most of those which have been discovered in such situations have been martins.

The instances of this sort are so numerous from all parts, that to bring them within a moderate compass I must only select a few of them; promising those who are incredulous, that I can most readily furnish many more than I shall now produce . . .

Mr. Manning, a surgeon of reputation in Kingsbridge, when a boy, and in search of sparrows nests, on a headland called the Hope, pulled out from under the thatch of an uninhabited house great numbers of swallows (or martins) which he considered as dead, but they afterwards revived; and their number amounted to more than 40. Mr. Manning recollects the fact at present as if it had been more recent, and likewise remembers, that the plumage was in perfect order; which was the case also with some martins, which I received myself during the winter, from Camerton in Somersetshire, in which there was not the least mark of putrefaction.

Another person drew out a great number of martins from the wall of an old castle in Wales during winter, and the heat of his hands recovered some of them so as to fly . . .

With regard to the third species of swallows, the sand martin, I have never been able to collect a decisive instance of their being observed at all during the winter . . . I will not therefore pretend to conjecture what may be their peculiar lurking places, though I conceive that they undoubtedly have such . . .

As for the fourth species, called the *Swift*, which is well known by its superior size, and being almost entirely black, Linnaeus asserts, that it winters in the holes of churches . . .

4 The Sin of Cruelty

Humphry Primatt, *The Duty of Mercy and the Sin of Cruelty to Brute Animals*, ed. Richard D. Ryder (1776 and 1992), Ch. Three.

Creatures at large God claims as his own peculiar property . . . The duty of men concerning animals that are wild by nature, lies in a very narrow compass: *let them alone*. Being God's property, and in his sight, God will provide for them. And it is enough for us that we invade not their province, but leave them unmolested and at liberty to perform the tasks, and answer the ends, for which God was pleased to create them. . . .

It is an instance of the wisdom and goodness of God, that the brutes should be animals irrational and dumb. As to *brutes of ferocity*, it is certain, that if, beside their strength, swiftness and sharpness of tooth or talon, they were endued with the powers of reason and speech, men, who are animals naturally defenceless, and comparatively slow of motion, would live in perpetual fear and dread of them. And with regard to *brutes of humanity*, particularly the large and laborious kind, were they capable of reason, the reflection upon their subordinate and servile condition would render them very unhappy in themselves; and perhaps less tractable, and consequently less useful to us. And if to the power of reason, we suppose them likewise endued with the power of speech, the inconvenience to men would be much greater. For these brutes, by the united faculties of reason and speech, would be able to enter into combinations and conspiracies against mankind . . .

It is a further proof of the goodness and providence of God, that the large brutes of humanity, whose great strength and stature we are so much indebted to for labour and draught, should be so remarkably *tractable* and *tame*. For if, with their strength and stature, they had that savageness and ferocity of heart . . . which many other large brutes have . . . we should as submissively bow down before their magnanimity and power, as we now insult over their timidity and inoffensiveness.

For our service, God has been pleased to create these useful animals large and strong; and for our security it is, that they are timid, irrational and dumb. But, certainly, it does not become us to take a cruel advantage of any of their incapacities or defects, which are only intended as the reins by which we are to guide and control them. They are tendered to us with strength sufficient for labour; but with hearts which are humbled, and mollified, and willing to submit to the delightful and noble service of being useful unto men. And happy are they, when they find we accept their willing obedience, by our kind and tender usage of them. They are entirely in our power, and committed to our care. And it is not improbable that God has assigned his own providence over them to us, that they may be the more tractable, the more they find themselves dependent upon us. But then it is our duty to consider that . . . their own natural incapacities, lay a kind of claim and demand upon our attention and tenderness. There is a condition and restriction implied in the compact . . . the power granted to men to rule over the brutes, cannot be a power to abuse or oppress them. . . .

Yet ungrateful man, with all his reason unreasonable, deaf to the voice of justice, and obdurate to the feelings of compassion, abuses his power and dominion over these poor creatures; because, for his sake, they are defenceless, irrational, and dumb; because they are unable to resist us, and have neither argument to convince us of our injustice, nor speech to utter their complaints . . .

5 'Design' in Shell-bearing Animals

William Paley, *Natural Theology, or Evidences of the Existence and Attributes of the Deity* (1802), Ch xix.

I must now crave the reader's permission to introduce into this place, for want of a better, an observation or two upon the tribe of animals, whether belonging to land or water, which are covered by *shells.*

The *shells* of *snails* are a wonderful, a mechanical, and, if one might so speak concerning the works of nature, an original contrivance . . . which can, with no probability, be referred to any other cause than to express design . . . whilst a snail, as it should seem, is the most numb and unprovided of all artificers . . .

The shell of a lobster's tail, in its articulations and overlappings, represents the jointed part of a coat of mail; or rather, which I believe to be the truth, a coat of mail is an imitation of a lobster's shell. The same end is to be answered by both: the same properties, therefore, are required in both, namely, hardness and flexibility, a covering which may guard the part without obstructing its motion. For this double purpose, the art of man, expressly exercised upon the subject, has not been able to devise anything better than what nature presents to his observation. Is not this therefore a mechanism, which the mechanic, having a similar purpose in view, adopts? Is the structure of a coat of mail to be referred to art? Is the same structure of the lobster, conducing to the same use, to be referred to any thing less than art?

Some, who may acknowledge the imitation, and assent to the inference which we draw from it, in the instance before us, may be disposed, possibly, to ask, why such imitations are not more frequent than they are, if it be true, as we alledge, that the same principle of intelligence, design, and mechanical contrivance, was exerted in the formation of natural bodies, as we employ in the making of the various instruments by which our purposes are served. The answers to this question are, first, that it seldom happens, that precisely the same purpose, and no other, is pursued in any work which we compare of nature and of art; secondly, that it still seldom happens, that we *can* imitate nature, if we would. Our materials and our workmanship are equally deficient. . . In the example which we have selected, I mean of a lobster's shell compared with a coat of mail, these difficulties stand less in the way, than in almost any other that can be assigned; and the consequence is, as we have seen, that art gladly borrows from nature her contrivance, and imitates it closely . . .

BIBLIOGRAPHY

Anon, 'Cruelty to Brutes an Offence Against God', *The Gentleman's Magazine*, vol. xxxi (1762): p. 201.

Anon, 'A review of "White's Natural History and Antiquities of Selborne"', *The Topographer, April* (1789): No. 1.

Anon, *Hints for the Management of Hot-beds, and Directions for the Culture of Early Cucumbers and Melons*. Bath, 1790.

Adanson, M. *Voyage to Senegal, the Isle of Goree, and the River Gambia*, 'translated from the French'. London, 1759.

Addison, J. *et al. The Spectator*. London 1711–14

Albin, E. *The Natural History of Birds*. London, 1731–38

Allen, D. E. *The Naturalist in Britain*. London: Allen Lane, 1976.

Baldwin, J. M. *Development and Evolution*. New York: Macmillan, 1902.

Banks, Sir J. *The Endeavour Journal of Joseph Banks*, ed. J. C. Beaglehole. Sydney: Angus and Robertson, 1963.

Barker, T. 'Extract of a Register of the Barometer, Thermometer, and Rain, at Lyndon in Rutland, 1775', *Philosophical Transactions*, vol. 67 (1776): p. 350.

Barnard, J. C. *Animal Behaviour; Ecology and Evolution*. London: Croom Helm, 1983.

Barrington, Hon. D. 'An Essay on the Periodical Appearing and Disappearing of Certain Birds, at Different Times of Year', *Philosophical Transactions,* vol. 62 (1772): p. 265.

Barrington, Hon. D. 'Experiments and Observations on the Singing of Birds', *Philosophical Transactions*, vol. 63 (1774): p. 249.

Barrington, Hon. D. *Miscellanies*. London, 1781.

Bartram, W. *The Travels of William Bartram,* ed. F. Harper. New Haven, Conn: Yale University Press, 1958.

Belon, P. *L'Histoire de la nature des oyseaux, avec leurs descriptions et naifs pourtraicts*. Paris, 1555.

Berkeley, G. *The Principles of Human Knowledge, and Three Dialogues Between Hylas and Philonous*. London, 1734.

Bewick, T. *A History of British Birds*. Newcastle, 1797.

Black, W. *An Arithmetical and Medical Analysis of the Diseases and the Mortality of the Human Species*. London, 1789.

Bonner, J. T. *The Evolution of Culture in Animals*. Princeton: Princeton University Press, 1980.

Borlase, W. *The Natural History of Cornwall*. Oxford, 1758.

Borlase, W. *The Antiquities, Historical and Monumental, of Cornwall*. London, 1769.

Boswell, J. *London Journal 1762-1763*, ed. F. A. Pottle. New Haven, Conn: Yale University Press, 1950.
Boyle, R. *The Usefulnesse of Experimental Naturall Philosophy*. London, 1683.
Boyle, R. *The Works of Robert Boyle*, ed. T. Birch, vols I-V. London, 1744.
Bridgeman, P. W. *The Logic of Modern Physics*. New York: Macmillan, 1948.
Brisson, M. J. *Ornithologia*. Paris, 1760.
Browne, Sir T. *Certain Miscellany Tracts*, ed. Thomas Tenison. London, 1683.
Buffon (Georges Louis Leclerc), *Histoire naturelle*. Paris, 1749-88. Including *Ornithologie historique,* 1770.
Burnett, J. *A History of the Cost of Living*. Harmondsworth: Penguin, 1989.
Burnet, T. *The Sacred Theory of the Earth*. London, 1681.
Butler, J. *Fifteen Sermons* (1726). London: Bell, 1953.
Butler, J. *The Analogy of Religion, Natural and Revealed, to the Constitution and Course of Nature*. London, 1736.
Butterfield, H. *The Whig Interpretation of History, Medicine and the Life Sciences*. London: Bell, 1931.
Butts R. E. and Davis, J. W. (eds), *The Methodological Heritage of Newton*. Oxford: Oxford University Press, 1970.
Caplan, A. L. (ed.), *The Sociobiology Debate*. New York: Harper and Row, 1978.
Catesby, M. *The Natural History of Carolina, Florida and the Bahama Islands*. London, 1731-34.
Catesby, M. 'On the Migration of Birds', *Philosophical Transactions*, vol. 44, (1747): p. 435. Reprinted in shortened form in *The Gentleman's Magazine*, vol. xviii (1748), p. 447.
Cobbett, W. *Rural Rides*. London: Waters, 1826.
Collinson, P. 'Concerning the Migration of Swallows', *Philosophical Transactions*, vol. 51 (1760): p. 459.
Cox, C. B. and Moore, P. D. *Biogeography*. Oxford: Oxford University Press, 1973.
Crebbs, J.R. and Davies, N.B. *Behavioural Ecology*. Oxford: Blackwell, 1997.
Darling, F. F. *A Herd of Red Deer*. Oxford: Oxford University Press, 1936.
Darwin, C. *The Origin of Species by Means of Natural Selection*. London: John Murray, 1859.
Darwin, C. *The Expression of the Emotions in Man and Animals*. London: John Murray, 1872.
Darwin, C. *The Formation of Vegetable Mould Through the Action of Worms*. London: John Murray, 1881.
Darwin, C. *The Origin of Species by Means of Natural Selection,* 6th edn. London: John Murray, 1886.
Darwin, E. *Zoonomia, or the Laws of Organic Life*. London, 1794-96.
Derham, W. 'Concerning the Migration of Birds', *Philosophical Transactions*,

vol. 26 (1712): p. 123.
Derham, W. *Physico-Theology; or a Demonstration of the Being and Attributes of God, from his Works of Creation*. London, 1712.
Elton, C. *The Pattern of Animal Communities*. London: Methuen, 1966.
Emden, C. S. *Gilbert White in his Village*. Oxford: Oxford University Press, 1956.
Endler, J. A. *Natural Selection in the Wild*. Princeton: Princeton University Press, 1986.
Fabre, J. H. *Souvenirs entomologiques*. Paris: Delagrave, 1879-1907.
Fereyabend, P. *Farewell to Reason*. London: Verso, 1987.
Fielding, H. *The History of Tom Jones* (1749). Harmondsworth: Penguin, 1966.
Fisher J. and Hinde R. A. 'The Opening of Milk Bottles by Birds', *British Birds*, vol. 42 (1949), p. 337.
Fisher, J. *Birds as Animals; I, A History of Birds*. London: Hutchinson, 1954.
Fisher, J. *Watching Birds*, revised Jim Flegg. Berkhamsted: Poyser, 1974.
Foster, P. G. M. 'The Gibraltar Correspondence of Gilbert White', *Notes and Queries*, vol. 32 (1985): pp. 227, 315 and 489.
Foster, P. G. M. *Gilbert White and his Records*. Bromley: Christopher Helm, 1988.
von Frisch, K. *The Dancing Bees*. London: Methuen, 1954.
Gauthreaux, S. A. *Animal Migration, Orientation and Navigation*. New York: Academic Press, 1980.
Geist, V. *Mountain Sheep: a Study in Behaviour and Evolution*. Chicago: University of Chigago, 1971.
Geoffroy, Etienne Louis. *Histoire abregee des insectes*, revised edn. Paris, 1764.
Gibbons, D. W., Reid, J. B. and Chapman, R. A. (eds), *New Atlas of Breeding Birds in Britain and Ireland*. Berkhamsted: Poyser, 1991.
Gibson, J. *Locke's Theory of Knowledge and its Historical Relations*. Cambridge: Cambridge University Press, 1960.
Gilbert L. E. and Raven, P. H. (eds), *Coevolution of Animals and Plants*. Austin: University of Texas Press, 1975.
Gilpin, W. *Dialogue upon the Gardens at Stowe in Buckinghamshire*. London, 1748.
Glass, H. B. (ed.), *Fore-runners of Darwin 1745-1859*. Baltimore: Johns Hopkins University Press, 1959.
Goodall, J. *In the Shadow of Man*. London: Collins, 1971.
Goodall, J. and van Lawick, H. *Innocent Killers*. London: Collins, 1970.
Gould, S. J. *Ever Since Darwin*. London: Burnett/Deutsch, 1978.
Gould, W. *An Account of English Ants*. London, 1747.
Gray, E. A. *Portrait of a Surgeon, a Biography of John Hunter*. London: Hale, 1952.
Greig-Smith, P. W. *The Behavioural Ecology of the Stonechat,* Saxicola torquata. Unpublished Ph.D. thesis, Oxford University, 1979.
Griffin, D. R. *Animal Thinking*. Cambridge, Mass: Harvard University Press,

1984.
Hales, S. 'Vegetable Statics' and 'Haemastaticks', *Statical Essays*, vols I and II. London, 1733.
Hall, A. R. *The Scientific Revolution, 1500-1800*. London: Longmans, 1954.
Hampson, N. *The Enlightenment*. Harmondsworth: Penguin, 1968.
Hardy, Sir A. C. *The Living Stream*. London: Collins, 1965.
Harris, M. *The Aurelian; or Natural History of English Insects, namely Moths and Butterflies*. London, 1766.
Harvey, P. H. and Pagel, M. D. *The Comparative Method in Evolutionary Biology*. Oxford: Oxford University Press, 1991.
Hasselquist, F. *Voyages and Travels in the Levant, 1749-1752*, 'translated into English'. London, 1766.
Henrey, B. *British Botanical and Horticultural Literature before 1800*, vols 2 and 3. Oxford: Oxford University Press, 1975.
Heyes, C. M. and Bennet Galef, G. (eds), *Social Learning in Animals*. London: Academic Press, 1996.
Hinde, R. A. (ed.), *Non-Verbal Communication*. Cambridge: Cambridge University Press, 1972.
Hinde, R. A. *Ethology: its Nature and Relations with Other Sciences*. London: Fontana, 1982.
Holt-White, R. *The Life and Letters of Gilbert White*. London: John Murray, 1901.
Hooke, R. *Micrographia*. London, 1665.
Hooke, R. *The Posthumous Works of Robert Hooke*, ed. R. Waller. London, 1705.
Howard, H. E. *Territory in Bird Life*. London: John Murray, 1920.
Hudson, W. *Flora Anglica*. London, 1762.
Hume, D. *A Treatise of Human Nature* (1739). Oxford: Oxford University Press, 1888.
Hume, D. *Dialogues Concerning Natural Religion*. London, 1779.
Hume, D. *The Letters of David Hume*, ed. J. Y. T. Grieg, vols I and II. Oxford: Oxford University Press, 1932.
Hunt, G. R. 'Manufacture and Use of Hook-tools by New Caledonian Crows', *Nature*, vol. 379 (1996): p. 249.
Hunter, J. *Observations on Certain Parts of the Animal Oeconomy*. London, 1786.
Hutton, J. *The Theory of the Earth, with Proofs and Illustrations*. Edinburgh, 1795.
Huxham, J. *Observationes de Aera, et Morbis Epidemicis, 1727-1748*. London, 1749.
Huxley, J. 'The Courtship Habits of the Great Crested Grebe', *Proceedings of the Zoological Society*, vol. 1914 (1915): p. 491.
Huxley, J. *Memories*, vol. II. London: Allen and Unwin, 1973.
Huxley, T. H. *Evidences as to Man's Place in Nature*. London: Williams and Norgate, 1863.
Jardine, N., Second J. A. and Spary, E. C. *Cultures of Natural History*. Cam-

bridge: Cambridge University Press, 1996.
Jenner, E. 'Observations on the Natural History of the Cuckoo'. *Philosophical Transactions*, vol. 78 (1788): p. 219.
Jenner, E. 'Some Observations on the Migration of Birds', *Philosophical Transactions,* vol. 114 (1824): p. 11.
Johnson, W. *Gilbert White, Pioneer, Poet and Stylist.* London: John Murray, 1928.
Kalm, P. *Travels into North America*, trans. J. R. Forster. Warrington, 1770-71.
Kamil A. C. and Jones, J. E. 'The Seed-storing Corvid Clark's Nutcracker Learns Geometric Relationships Among Landmarks'. *Nature*, vol. 390 (1997): p. 236.
Kirby, W. *The Power, Wisdom and Goodness of God as Manifested in the Creation of Animals.* (Bridgewater Treatises No. 7). London, 1835.
Knight, D. M. *The Nature of Science; the History of Science in Western Culture Since 1600.* London: Andre Deutsch, 1976.
Koehler, Otto. 'Non-Verbal Thinking', in Heinz Friedrich, (ed.), *Man and Animal.* London: Granada, 1971.
Kuhn, T. S. *The Structure of Scientific Revolutions*, revised edn. Chicago: University of Chicago Press, 1970.
Lack, D. *The Life of the Robin.* London: Witherby, 1943.
Lack, D. *Darwin's Finches.* Cambridge: Cambridge University Press, 1947.
Lack, D. *Ecological Adaptation for Breeding in Birds.* London: Methuen, 1968.
de Lamarck, J. B. *Philosophie Zoologique*. Paris, 1809.
Legg, J. *The Emigration of British Birds.* Salisbury, 1780.
Lendrem, D. *Modelling in Behavioural Ecology.* London: Croom Helm, 1986.
Leroy, C. G. *Lettres philosophique sur l'Intelligence et la perfectibilite des Animaux.* Paris, 1764.
Levins R. and Lewontin, R. C. *Dialectical Biology.* Cambridge, Mass: Harvard University Press, 1985.
Lewis, C. I. *Mind and the World-Order,* revised edn. New York: Dover, 1956.
Lewontin, R. C. *The Genetic Basis of Evolutionary Change.* New York: Columbia University Press, 1974.
Linnaeus (Carl von Linne), *Systema Naturae.* Stockholm, 1735.
Linnaeus (Carl von Linne), *Systema naturae*, 10th edn. Stockholm, 1758.
Linnaeus (Carl von Linne), *An Introduction to Botany: Extracts from Linnaeus*, ed. J. Lee, 2nd edn. London, 1765.
Locke, J. *An Essay Concerning Human Understanding* (1690). London: Dent, Dutton, 1961.
Lockley, R. M. *A Biography of Gilbert White.* London: Witherby, 1954.
Lockley, R. M. *The Island.* London: Andre deutsch, 1969.
Lorenz, K. *King Solomon's Ring*. London: Methuen, 1952.
Lorenz, K. 'The Companion in the Bird's World', *Auk*, vol. 54 (1937): p.

245.
Lorenz, K. 'Pair-Formation in Ravens', in Heinz Friedrich, (ed.), *Man and Animal*. London: Granada, 1971.
Lorenz, K. 'The Comparative Study of Behaviour', in Konrad Lorenz and Paul Leyhausen, (eds), *Motivation of Human and Animal Behaviour*. New York: Nostrand Reinhold, 1973.
Lyell, C. *The Principles of Geology*. London: John Murray, 1830-33.
Mabey, R. *Gilbert White, a Biography*. London: Century Hutchinson, 1986.
McAlpine, D. *The Botanical Atlas, a Guide to the Practical Study of Plants*. Edinburgh: Johnson, 1883.
Maclean, K. *John Locke and English Literature of the Eighteenth Century*. New Haven, Conn: Yale University Press, 1936.
Malthus, T. *An Essay on the Principles of Population*. London: Johnson, 1798.
Manning A. and Stamp Dawkins, M. *An Introduction to Animal Behaviour*, revised edn. Cambridge: Cambridge University Press, 1993.
Marshall, W. *Minutes of Agriculture, Made on a Farm of 300 Acres of Various Soils*. London, 1787.
Martin, B. *A Plain and Familiar Introduction to the Newtonian Philosophy*. London, 1751.
Maupertuis, P. L. *Systeme de la nature*. Paris, 1751.
Maynard Smith, J. *The Problems of Biology*. Oxford: Oxford University Press, 1986.
Maynard Smith, J. and Szathmary E. *The Origins of Life*. Oxford: Oxford University Press, 1999.
Mayr, E. *Animal Species and Evolution*. Cambridge, Mass: Harvard University Press, 1963.
Mayr, E. *The Growth of Biological Thought* Cambridge, Mass: Harvard University Press, 1982.
Millar, J. *Observations on the Practice in the Medical Department of the Westminster General Dispensary*. London, 1777.
Miller, P. *The Gardener's Dictionary*, 6th edn. London, 1753.
Montagu, G. *The Ornithological Dictionary*. London, 1802.
Morgan, C. L. *Animal Behaviour*. London: Arnold, 1900.
Morton, J. *The Natural History of Northampton-shire*. London, 1712.
Mullet, C. '*Multum in Parvo*: Gilbert White of Selborne', *Journal of the History of Biology*, vol. 2 (1969): p. 363.
Newton, I. *Opticks; or a Treatise of Light*. London, 1704.
Newton, I. *The Correspondence of Isaac Newton*, ed. W. H. Turnbull, vol. 1. Cambridge: Cambridge University Press, 1959.
Nicholson, E. M. *How Birds Live*, 2nd edn. London: Williams and Norgate, 1929.
Ormerod, S. J. 'Pre-Migratory and Migratory Movements of Swallows *Hirundo rustica* in Britain and Ireland', *Bird Study*, vol. 38 (1991): p. 170.
Paley, W. *Natural Theology; or Evidences of the Existence and Attributes of the Deity*. London, 1803.

Passmore, J. A. *Hume's Intentions*, 3rd edn. London: Duckworth, 1980.
Patten, B. C. (ed.) *Systems Analysis and Simulation in Ecology*, 1 and 2. London: Academic Press, 1971–73.
Patterson, F. and Linden. E. *The Education of Koko*. London: Andre Deutsch, 1982.
Pennant, T. *British Zoology,* 4th edn. London, 1776.
Pepperberg, I. 'Proficient Performance of a Conjunctive, Recursive Task by an African Gray Parrot', *Journal of Comparative Psychology*, No. 106 (1992): p. 295.
Piaget, J. *Behaviour and Evolution.* London: Routledge and Kegan Paul, 1979.
Pickstock, J. C., Krebs J. R. and Bradbury, S. 'Quantitative Comparison of Sonograms Using an Automatic Image Analyser. Application to Song Dialects of Chaffinches, *Fringilla coelebs*', *Ibis*, No. 122 (1980): p. 103.
Plot, R. *The Natural History of Oxfordshire*. Oxford, 1676.
Plot, R. *The Natural History of Staffordshire*. Oxford, 1686.
Plotkin, H. C. (ed.), *The Role of Behaviour in Evolution.* Cambridge, Mass: Harvard University Press, 1988.
Pluche, N. A. *Spectacle de la nature*. Paris, 1732-50.
Pope, A. 'Against Barbarity to Animals', *The Guardian*, No. lxi (1713): May 21.
Pope, A. *An Essay on Man.* London, 1733-34.
Pope, A. 'Of the Knowledge and Characters of Men' and 'Of the Use of Riches', *Moral Essays*. London, 1738.
Primatt, H. *The Duty of Mercy and Sin of Cruelty to Brute Animals, 1776,* ed. Richard D. Ryder. Fontwell: Centaur, 1992.
Prochaska, G. *Dissertation on the Functions of the Nervous System.* London, 1784.
Raven, C. E. *John Ray, Naturalist: his Life and Works*. Cambridge: Cambridge University Press, 1942.
Ray, J. *Methodus Plantarum Nova*. London, 1682.
Ray, J. *The Wisdom of God Manifested in the Works of the Creation.* London, 1691.
Reaumur (Rene Antoine Ferchault). *Memoires pour servir a L'histoire naturelle des insects*. Paris, 1734-42.
Ritchie, A. D. *George Berkeley: a Reappraisal.* Manchester: Manchester University Press, 1967.
Rogers, G. A. J. (ed.), *Locke's Philosophy*. Oxford: Oxford University Press, 1994.
Rousseau, C. S. 'Science', in Rogers, P. (ed.), *The Eighteenth Century*. London: Methuen, 1978.
Schofield, R. E. *Mechanism and Materialism: British Natural Philosophy in the Age of Reason*. Princeton: Princeton University Press, 1970.
Scopoli, J. A. *Annus Primus Historico-Naturalis*. Leipzig, 1769.
Selous, E. *Realities of Bird Life*. London: Constable, 1927.
Shapin, S. and Schaffer, S. *Leviathan and the Air-pump*. Princeton: Princeton

University Press, 1985.

Slater, P. J. B. and Halliday, T. R. (eds), *Behaviour and Evolution.* Cambridge: Cambridge University Press, 1994.

Smollett, T. *The Expedition of Humphry Clinker* (1771). Harmondsworth: Penguin, 1985.

Spalding, D. 'Instinct, with Original Observations on Young Animals', *Macmillan's Magazine,* vol. 27 (1877): p. 282.

Stace, C. A. *New Flora of the British Isles.* Cambridge: Cambridge University Press, 1991.

Stamp Dawkins, M. *Through Our Eyes Only? The Search for Animal Consciousness.* Oxford: Freeman, 1993.

Stebbing, L.S. *Philosophy and the Physicists.* Harmondsworth: Penguin, 1944.

Sterne, T. *A Sentimental Journey*, London, 1766.

Stillingfleet, B. *Miscellaneous Tracts Relating to Natural History, Husbandry and Physick,* 2nd edn. London, 1762.

Swammerdam, J. *The Book of Nature and the History of Insects,* trans Thomas Floyd. London, 1758.

Swift, J. *A Tale of a Tub and The Battle of the Books* (1704). London: Dent, Dutton, 1909.

Tansley, A. G. *The British Isles and their Vegetation.* Cambridge: Cambridge University Press, 1939.

Teich, M. and Young, R. (eds), *Changing Perspectives in the History of Science.* London: Heinemann, 1973.

Thomas, K. *Man and the Natural World: Changing Attitudes in England 1500-1800.* London: Allen Lane, 1983.

Thomson, J. *The Seasons.* London, 1726-30.

Thorpe, W. H. *Learning and Instinct in Animals.* London: Heinemann, 1956.

Thorpe, W. H. 'The Learning of Song Patterns by Birds, with Special Reference to the Song of the Chaffinch', *Ibis*, No. 109 (1958): p. 535.

Thorpe, W. H. *The Origins and Rise of Ethology.* London: Heinemann, 1979.

Tinbergen, N. *The Herring Gull's World.* London: Collins, 1953.

Tinbergen, N. *Social Behaviour in Animals.* London: Methuen, 1965.

Tinbergen, N. *The Study of Instinct,* revised edn. Oxford: Oxford University Press, 1969.

Trohler, T. 'To Improve the Evidence of Medicine', *History and Philosophy of the Life Sciences*, vol. 10 (1988): Supplement, p. 31.

Tull, Jethro. *The Horse-Hoing Husbandry: or, an Essay on the Principles of Tillage and Vegetation.* London, 1733.

Turner, R. *Capability Brown and the Eighteenth-Century English Landscape.* London: Weidenfeld and Nicolson, 1985.

Turrill, W. B. *British Plant Life,* 3rd edn. London: Collins, 1962.

Unwin, D. M. *Microclimate Measurement for Ecologists.* London: Academic Press, 1980.

Virgil, *The Georgics*, trans. C. Day Lewis. London: Jonathan Cape, 1943.

Waddington, C. H. *The Evolution of an Evolutionist.* Edinburgh: Edinburgh

University Press, 1975.

Waterton, C. *Essays on Natural History*. London, 1838-57.

Watson, J. B. *Psychology from the Standpoint of a Behaviourist*, 2nd edn. Philadelphia: University of Pennsylvania Press, 1924.

Watson, R. *Chemical Essays*, vol. III. London, 1785.

White, G. 'Of the House-martin, or Martlet', *Philosophical Transactions*, vol. 64 (1774): p. 196.

White, G. 'Of the House-Swallow, the Swift or Black Martin, and the Sand-Martin', *Philosophical Transactions*, vol. 65 (1775): p. 258.

White, G. *Garden Kalendar, 1751-1771*, facsimile, ed. John Clegg. London: Scolar Press, 1975.

White, G. *The Journals of Gilbert White, 1751-1793,* vols I-III, ed. Francesca Greenoak. London: Century Hutchinson, 1986-89.

White, G. *The Natural History and Antiquities of Selborne, in the County of Southampton*. London, 1789.

White, G. *The Natural History and Antiquities of Selborne,* ed. J. Mitford. London, 1813.

White, G. *The Natural History and Antiquities of Selborne*, ed. E. Jesse. London: Bohn, 1851.

White, G. *The Natural History and Antiquities of Selborne,* ed. Thomas Bell. London: van Voorst, 1877.

White, G. *The Natural History and Antiquities of Selborne and A Garden Kalendar*, ed. R. Bowdler Sharpe. London: Freemantle, 1900.

White, G. *The Natural History of Selborne,* ed. E. M. Nicholson. London: Thornton Butterworth, 1929.

White, G. *The Writings of Gilbert White of Selborne,* ed. H. J. Massingham. London: Nonesuch Press, 1938.

White, G. *The Natural History of Selborne*, ed. James Fisher. London: Cresset Press, 1947.

White, G. *The Natural History of Selborne,* ed. Richard Mabey. Harmondsworth: Penguin, 1977.

White, T. 'Remarks on Oaks', *Gentleman's Magazine*, vol. lv (1785): p. 109.

Whitehead, A. N. *Science and the Modern World*. Cambridge: Cambridge University Press, 1926.

Whiten, A. *et al.* 'Cultures in Chimpanzees'. *Nature*, vol. 399 (1999): p. 682.

Wildman, T. *A Treatise on the Management of Bees*. London, 1768.

Willey, B. *The Eighteenth Century Background*. London: Chatto and Windus, 1940.

Willughby, F. *The Ornithology of Francis Willughby*, trans. and ed. John Ray. London, 1678.

Woodward, J. *Essay Towards a Natural History of the Earth*. London, 1695.

Young, A. *A Six Weeks' Tour Through the Southern Counties of England and Wales*. London, 1768.

NOTES AND REFERENCES

TP *Natural History of Selborne*, letters to Thomas Pennant.

DB *Natural History of Selborne*, letters to Daines Barrington.

AN *Antiquities*.

NJ *Naturalist's Journal*.

GC *Garden Calendar*.

PF Paul G. M. Foster, 'The Gibraltar Correspondence of Gilbert White', *Notes and Queries*, vol. 32 (1985): pp. 227–36, 315–28 and 489–500.

RH Rashleigh Holt-White, *The Life and Letters of Gilbert White* (1901).

WJ Walter Johnson, *Gilbert White; Pioneer, Poet and Stylist* (1928).

The British Library catalogue numbers for the *Naturalist's Journal* are MS Add: 31,846–31,851. For quotations from *Selborne* and the *Antiquities*, the corrected edition of 1813 has been used throughout. In quotations from the MS journals and notebooks, the spelling, punctuation, use of capitals and so on in the MS have generally been retained, although in references to the *Naturalist's Journal*, in particular, there is sometimes room for disagreement as to just which date should be attached to a given entry.

Preface

1 *Gardeners' Question Time*, BBC Radio Four, Feb. 12, 1984.
2 DB xxxv.
3 H. R. Hays, *Birds, Beasts and Men* (1973), Ch. 9; Walter S. Scott, *White of Selborne and His Times* (1946), Ch. Five; and David E. Allen, *The Naturalist in Britain* (1979), Ch. Two.
4 This story could be repeated even by Ronald Lockley. See *The Natural History of Selborne*, ed. R. M. Lockley (1949), Introduction.
5 *Man and the Natural World*, 1500–1800 (1983), VI : iii.
6 The 'mythology' oddly complicates the discussion of Gilbert White. For example, the only undoubtedly authentic portraits of the naturalist are two small pen drawings, added by a certain 'T.C.' to White's copy of Pope's translation of the *Iliad*. They show us a slight, beaky-nosed man, who

looks – for what this is worth – sensitive and alert. But for some years in the mid-twentieth century the painting of a heavier, almost stolid-looking gentleman was also claimed as a Gilbert White portrait: this no doubt was the 'contented' or 'static' White. See, for example, Walter S. Scott, *White of Selborne and His Times* (1946); and for the contrasting portraits placed side by side, R. M. Lockley, *A Biography of Gilbert White* (1954)).

7 PF pp. 322, 323.

8 Four letters written to White by the Oxford botanist William Sheffield give the same impression, even though we do not have White's contribution to this exchange. See Ch. 9, note 29, below.

9 PF p. 492. Where White gets most embroiled in classificatory problems, in one of the last of the Gibraltar letters, we most sense his impatience with the fine detail of classification. PF p. 495.

10 *Memories* (1973), vol. II, Ch. xvii.

11 *A Herd of Red Deer* (1936), Ch. 11.

12 *The Works of George Berkeley*, ed. H. A. Luce and T. E. Jessop (1948), vol. One, Notebook B.

13 A complete transcription of White's journals is now available. *The Journals of Gilbert White, 1751–1793*, has been produced by Francesca Greenoak, under the general editorship of Richard Mabey. vol. I (1986), covers the *Garden Calendar*, the *Floral Calendar* and the *Naturalist's Journal* up to the year 1773. vol. II (1988), covers the *Naturalist's Journal* from 1774 to 1783; and vol. III (1989), the *Naturalist's Journal* from 1784 to June 1793, the month of White's death. The work has been meticulously annotated, and without doubt provides a valuable alternative to scrutinising the relevant microfilm. Although, with the microfilm or the manuscript itself in mind, one caveat can be suggested: White's positioning of material on the manuscript pages is often of considerable interest, and this is, of course, partly lost in the process of adapting the material for letterpress.

14 See eg H. J. Massingham, ed, *The Writings of Gilbert White of Selborne,* Nonesuch Press, 1938, Introduction.

15 G. Kitson Clark, *The Making of Victorian England* (1962), Ch. 1. Wisely, as it can now seem, Kitson Clark would not be quite dominated by 'culture patterns'.

16 *The Natural History of Selborne* (1929), Introduction.

17 WJ Ch. Two.

18 *Watching Birds* (1946), Ch. III.

19 WJ Ch. Nine.

20 The recently revived notion that a behavioural emphasis necessarily *leads to* the depiction of living things as machine-like, is mistaken.

21 At numerous points, he is vindicated in the work of 'dynamic' and even evolutionary naturalists. See Ch. 10, below.

22 Johnson's study, published originally by John Murray (1928), has been re-issued under various imprints, most recently by Macdonald-Futura (1981), as *Gilbert White*. A facsimile of White's first journal, *Gilbert*

White, Garden Kalendar, 1751–1771, was produced by Scolar Press (1975), and is of great value to students. Francesca Greenoak's complete transcription of the diaries and journals has already been referred to (see note 12 above). The spelling 'kalendar', incidentally, was abandoned by White towards the end of his first journal in favour of 'calendar', and was not used again by him; for convenience, I have used 'calendar' throughout.

Chapter 1: Selborne and Natural History

1 TP x.
2 James Fisher, *The Birds of Britain* (1942), p. 38.
3 When White's correspondent Thomas Pennant was due to send a packing case back to him from Wales, White wondered whether Pennant might like to include some Welsh fossils and rock samples with this container; but he added, 'a few will be sufficient'. RH vol. I, p. 207.
4 Konrad Lorenz, *King Solomon's Ring* (1952), Ch. 11.
5 DB xli.
6 TP x.
7 TP xi, and DB lvii.
8 See DB i (the 'white' is the pied wagtail), and, for example, NJ 1773, Dec. 25.
9 NJ 1789, July 5.
10 For 'real snipe's eggs', see NJ 1791, June 12.
11 Even allowing for the new *British Zoology* by Thomas Pennant (1761), the first quite reliable bird illustrations were those of Thomas Bewick, whose *History of British Birds* appeared only in 1797. When White began work, Willughby's *Ornithology*, which had been published first in 1676, and in John Ray's translation in 1678, was still a standard guide. As to the 'expensive, coloured' books; Moses Harris's *Aurelian, or Natural History of English Moths and Butterflies* (1766) appeared in a second edition in 1778 and sold for ten guineas or more, unbound, and Elisabeth Blackwell's widely admired *Curious Herbal*, published first in 1737 in two volumes, was selling by the second half of the century for six or seven guineas and on larger paper for ten guineas. During these years, an octavo-sized 'trade book' could be issued at between five and, say, eight shillings. For White's income, and other wages and prices, see Ch. 1, note 45, and Ch. 2, note 5, below.
12 Although White did not in fact find the eggs which set him on the right road here himself, he says of them; 'The eggs were oblong, dusky, & streaked somewhat in the manner of the plumage of the parent-bird, & were equal in size at each end. The dam was sitting on the eggs when found . . .' NJ 1789, July 14.
13 Respectively, *Phylloscopus collybita, trochilus* and *sibilatrix*. Today the last, the wood warbler, is 'widespread' but is also a rare 'summer bird' in Britain. See Ch. 5, below.

14 DB xlii. In this passage, White was responding directly to the view expressed by, for instance, his correspondent Daines Barrington. According to the latter: 'It is an old saying that "a bird in the hand is worth two in the bush", and this holds equally with regard to their being distinguished, when those even who study natural history, have but a transient sight of the animal.' 'On the Periodical Appearing and Disappearing of Birds, at Different Times of the Year', *Philosophical Transactions*, vol. 62 (1772), p. 265.
15 DB xv.
16 TP x.
17 'Autumn', lines 669–671; *The Seasons* (1726–30).
18 TP x and D.B v.
19 See, for example, DB ix and TP xiii.
20 In the 1776 edition of the *British Zoology,* though, Pennant did credit White explicitly with the discovery of the harvest mouse.
21 They were to become DB xvi, xviii, xx and xxi. For these papers as they appeared in the *Philosophical Transactions*, see Bibliography.
22 See R. M. Lockley, *A Biography of Gilbert White* (1954), Ch. iv.
23 GC but see also RH vol. I.
24 Even at this time some of the ground was still not owned outright, by himself or his family.
25 What I am calling the 'back' here was the original front of the house, but as we can think of it, the house now had two 'fronts'. (In a letter written in 1773, White refers to the part of the house facing the Hanger as 'my back front'.)
26 GC, and RH vol. I.
27 NJ 1777, June 6; 'Began to build the walls of my parlor, which is 23 feet & half by 18 feet; & 12 feet high & 3 inch'. White enlarged the house, but its present grand proportions are the result of nineteenth century extensions.
28 *Gilbert White, Garden Kalendar, 1751–1771*, ed. John Clegg (1975), Introduction.
29 GC 1759, Nov. 15.
30 GC 1759, Oct. 24.
31 The reader may wonder whether he does not mean barrow-loads, but these were 'dung carts-ful', brought in mainly from neighbouring farms. See also Ch. 2, below.
32 Regarding greenhouses, he notes in the first year of the *Naturalist's Journal*, 1768, Jan. 21, 'First crop of kidney-beans gather'd in the Hothouse at Hartley'; and on Dec. 22 of the same year, 'French beans are planted in the hot-house at Hartley. Pines are still cutting'. Hartley Mauditt lies two miles to the north of Selborne, and pineapples were still a curiosity. See also NJ 1769, Dec. 18.
33 *Garden Calendar;* and James Boswell, *London Journal, 1762–1763*, ed. Frederick A. Pottle (1950), December 15, 1762.
34 He bought an edition of Miller's *Gardener's Dictionary* in 1747, and

seems to have added another in 1753, when he also acquired John Ray's *Methodus Plantarum Nova*. See WJ Ch. Eight.

35 GC 1759, Feb. 17.

36 NJ 1787. But these figures too could be exceeded. During the 1780s, he began to give ridge, or outdoor, cucumbers serious attention, and during the late summer and early autumn of 1791 he and his assistants cut well over a thousand ridge cucumbers: NJ 1791. Many of them were no doubt given away or bartered, and his periodic references to 'pickling' must throw some light on this huge production. See Ch. 2, sect. i, below.

37 See Philip Miller, *The Gardener's Dictionary* (1753).

38 Had it not been for new and improved methods of land-management, and in particular of corn-growing, he remarks in retrospect, the nation at large would have suffered a famine during this rainy period. D.B xix. Selborne weather in White's time is discussed in some detail in Ch. 9, sect. iii, below.

39 TP i.

40 GC 1765, July 13.

41 In June 1782, he records, 'We have had no rain since June 13. The ground is bound as hard as iron, & chopped & cracked . . . The wetter the spring is, the more our ground binds in summer'. NJ 1782, June 30. A fortnight before he died, in June 1793, 'The ground all as hard as iron: we can sow nothing nor plant out'. NJ 1793, June 11. See also NJ 1777, Dec. 20.

42 Again the quantities can be noticed. In September 1754, '15 loads of melon-earth from Dorton, 8 shillings'. Account book, quoted in *The Natural History of Selborne*, ed. Thomas Bell (1877), vol. II, notes. Regarding the use of peat ashes, 'Sowed a good coat of ashes on Baker's hill, & also on the great meadow. Bought 40 bushels of ashes of Mrs Etty, & 36 bushels of sundry others. Sowed my own also.' NJ 1786, Feb. 4. A bushel was a volume measure equivalent to eight gallons.

43 GC 1765, April 12. Timothy Turner's ground was the same as White's, whereas Andrew Etty, the then vicar in residence at Selborne, enjoyed some of the 'good gardening soil' on the other side of the village street. Peat was an important domestic fuel in the parish, and 'peat dust' was the detritus from stored peat.

44 The early nineteenth century assumption that White would have been comfortably off *without* his Oxford fellowship was mistaken, but his other assets can also be under-estimated. As well as the fellowship, worth £100 annually, and the Faringdon salary, he had two other small stipends, a patrimony of £66 a year and some property rents. In the 1770s, his total income was perhaps £230 a year. This was what he would have received from a middling vicarage in the second half of the century; however, ordained Anglicans commonly 'pluralised' or enjoyed incomes from several different livings at once. See RH vol. I, pp. 104–8, and vol. II, pp. 44–48. Also R. M. Lockley, *A Biography of Gilbert White* (1954), and Richard Mabey, *Gilbert White, a Biography* (1986).

45 *The Georgics,* trans. C. Day Lewis, (1943), Bk. One, lines 50–56.

46 See Ch. 7, sect. i, below.
47 GC 1763, July 28.
48 GC 1758, March 25.

Chapter 2: Nature as an 'Economy'

1 A sixth brother, Francis, died at the age of 21. See R. M. Lockley, *A Biography of Gilbert White* (1954), Ch. ii.
2 For a discussion of John White's natural history, or 'Fauna Calpensis', see Paul G. M. Foster, *Gilbert White and his Records* (1988), Ch. Ten.
3 TP v and footnote. The total population of the parish in the 1780s was 'upwards of six hundred and seventy' people, with baptisms exceeding deaths.
4 NJ 1782, Oct. 13.
5 What the women did depended on the size of their families; they hoed the arable fields and helped at harvest time, or more regularly, worked as dairy maids; after a period of decline, the wool spinning which had been a cottage industry in the parish was 'a little revived'. TP v. Notes on prices and wages in White's time and locality are scattered through his journals and letters. Beef and mutton were usually 2d a pound, and after an exceptionally long and cold winter might rise to 4d or 5d a pound; in October 1781, after a drought period, White was paying an 'unprecedented' 9d a pound for butter. In the 1770s, a skilled carpenter could earn one shilling and tenpence a day, at Selborne, and a bricklayer and his assistant three shillings and sixpence a day between them. (During a good hop-picking, 'a woman and her girl' might earn two shillings a day or more. RH II, Ch. 2, and NJ 1786, Aug. 30. For a poor hop-picking, see NJ 1787, Sept. 10.) In London, a printer's compositor was paid three shillings and sixpence a day at this time, and Boswell thought a shilling for his dinner by no means unreasonable; but even considering cash wages alone, the Selborne villagers were well off compared with their counterparts in a number of more westerly counties. See John Burnett, *A History of the Cost of Living* (1969), Ch. 3.
6 See C. S. Emden, *Gilbert White in His Village* (1956). Also R. M. Lockley, *A Biography of Gilbert White* (1954) and Richard Mabey, *Gilbert White, a Biography* (1986).
7 NJ 1787, Nov. 16.
8 DB xxvi.
9 See, for example, NJ 1784, May 14; and TP ix (last paragraph). Cf. NJ 1789, April 19.
10 NJ 1789, Aug. 9, and TP xxii. He had long been intrigued by the nightjar, but in a letter dated Sept. 1, 1789, he had 'just found out' about the alleged responsibility of the nightjar for the damage to calves.
11 On one occasion, to help move a small barn bodily across one of his fields, he mustered '20 hands': GC 1766, Apr. 26.
12 Extras could be brought in from outside, of course. Brandy was purchased

by White regularly and port occasionally; from time to time he asked for some stockfish or a special ham to be sent down to him from London. For details, see RH vol. II.

13 We have no evidence of selling which, if selling took place, would be odd in someone who kept records and accounts meticulously. At South Lambeth, though, adjacent as it was to London, Benjamin White marketed produce regularly from his large gardens.

14 See *Garden Calendar*.

15 E.g. AN ii.

16 See also Ch. 8, below.

17 NJ 1789.

18 NJ 1792, Oct. 7. For hardship in a wet year, see also RH vol. II, p. 118.

19 TP viii.

20 NJ 1776, Aug. 26.

21 DB xxxv.

22 *The Formation of Vegetable Mould Through the Action of Worms* appeared in 1881.

23 DB xxix.

24 NJ 1781, Aug. 11, 24 and Oct. 14. See also DB lxiv.

25 NJ 1781, Oct. 31.

26 NJ 1781, Oct. 25 and Nov. 3. The importance of acorns in a good acorn year appears from numerous entries: in October 1783, a smallholder in the village – Timothy Turner again – bought 'upwards of 40 bushels' of acorns, NJ 1783, Oct. 23. The price for acorns was a shilling a bushel throughout the *Naturalist's Journal* years.

27 NJ 1781, Sept. 15 and TP i; also NJ 1781, Oct. 12 and 23.

28 See DB xxix. Cf. NJ 1778, Dec. 9, and 1783, June 21.

29 NJ 1776, June 10. See also NJ 1775, July 14 or 1785, Aug. 25.

30 DB xxix. See also AN xxv. The Kalm he refers to was Pehr Kalm, whose *Travels into North America* had been translated in 1770.

31 NJ 1782, May 31.

32 DB xx.

33 DB viii.

34 DB lxiv. Though a *Journal* entry dated Sept. 30, 1782, throws some doubt on 'only in hot summers' for wasps.

35 NJ 1781, Oct, 21.

36 NJ 1775, Dec. 1.

37 For the authoritative mid-twentieth century re-appraisal of John Ray, see C. E. Raven, *John Ray, Naturalist* (1942).

38 *The Analogy of Religion* (1736), I : vii : 4. Butler was an Anglican prelate, but in Hume's *Dialogues Concerning Natural Religion* (1779), the natural world is found to be more a 'great animal' or 'great vegetable' than a 'great machine'. Alexander Pope, in his *Essay on Man* (1733–34), refers to a 'Chain' of created beings, but also to a 'plastic' natural world; in this world, 'parts relate to whole'. *An Essay on Man, III.*

39 NJ 1789, July 19–21.

Chapter 3: The 'Outdoor' Method

1 E.g. DB ix.
2 DB x. The 'synonyms' were terms from several, varying taxonomical systems.
3 TP xxxi; see also DB vii. The quotation itself appeared on the title pages of early editions of *Selborne*. Carniola is part of what is now southern Austria.
4 DB xvi.
5 DB vii and xxii.
6 TP xxi.
7 A small, low-powered telescope, without a tripod, is entirely manageable as a bird-watching aid, but there is no mention of a telescope used for bird or animal observation in any of White's extant writings. He says he viewed Blanchard's balloon through a telescope, NJ 1784, Oct. 16, and according to Walter Johnson, a telescope must have been used in some of his astronomical sightings; although, as Johnson also points out, these sightings took place while Gilbert was staying with a sister and brother-in-law. WJ Ch. Two.
8 DB xxi. In this letter, the original dated 1774, he remarks that the young of swifts seem not to be fed by their parents in flight; but he kept the matter in mind. On July 20, 1784, he noted in the *Journal*, 'Saw an old Swift feed it's young in the air: a circumstance which I could never discover before'.
9 DB xv.
10 DB iii. When he had separated out the three leaf warblers, he was able to tell Pennant that they '*constantly* and *invariably* use distinct notes'. TP xix, White's italics.
11 WJ Ch. Nine.
12 TP xx and xxii. Pennant 'found', indeed, that even ring ouzels which bred in Scotland also wintered there; see TP xxvi. (By 'birds of passage', we are to understand any birds which seem to be migratory. 'Passage migrant' did not have the connotation for White that it has today.)
13 TP xxv.
14 TP xxxi. See also RH vol. I, pp. 252, 253.
15 TP xxxi.
16 DB vii. In his next letter to Barrington, DB viii, we notice, he quotes G. H. Kramer on the woodcock as a passage migrant in Austria.
17 DB xliv and DB xi. On the ancestry of the domestic dove, he again anticipates Charles Darwin. See *The Origin of Species* (1859), Ch. 1.
18 DB xxx. This enquiry seems to have arisen also from questions asked by Samuel Barker, one of White's nephews.
19 PF p. 323. See also notes on the Gibraltar correspondence in my Preface, above.
20 NJ 1784, June 5, 6 and Sept. 1. See also DB lxvi.
21 DB ii.

22 Second thoughts appear in the same way in his preface to the hirundine papers, DB xv. See Ch. 5, sect. iv, below.

23 Regarding 'rigorous experiment' in Britain, the eighteenth century often lagged behind the later seventeenth: see, for example, the discussion of Robert Boyle (1627–1691) in Steven Shapin and Simon Schaffer, *Leviathan and the Air-pump* (1985). White nowhere refers to these earlier experimentalists, but it is perhaps worth noticing that the first complete edition of Boyle's writings appeared in 1744. For various ideas concerning the absence – and in the case of 'medical arithmetic', the presence – of controls in eighteenth century science, I am indebted to Professor David Knight of Durham University.

24 PF p. 492.

25 By the 'condensed, selective' version of the *Naturalist's Journal*, I mean that of Walter Johnson (1931): admirable though this was in many ways, it entirely omitted the daily meteorological readings. For the new transcription of the *Journal*, by Francesca Greenoak, see Preface, note 13, above, and Bibliography. (The imprecise *Nature Calendar,* included in various nineteenth century editions of *Selborne,* and sometimes attributed to White, was in fact a selection from White's then unpublished writings made by James Aikin; it appeared first in 1801.)

26 Ronald Lockley can refer to the *Journal* as 'a phenological poem'.

27 Thus a note dated May 7, 1781 reads; 'Vast bloom among the apples: the crop but small'. And to a worried entry written on May 19, 1774, recording that the first leaves of his peaches and nectarines were 'sadly blotched & rivelled', is added, 'They bore fine fruit in plenty, considering the wet shady season'.

28 For the history of this 'diary blank', see Paul G. M. Foster, *Gilbert White and his Records* (1988), Ch. Eight.

29 NJ 1788, May 15.

30 NJ 1790, Aug. 16.

31 Occasionally, White wrote up his *Journal* while he was away from home; although the met figures, added when he returned, were always those for Selborne (they had been kept by Thomas Hoar). Other Selborne news, horticultural, ornithological or human, could be added where relevant.

32 TP xv.

33 NJ 1781, Dec. 9. George Tanner was the village shoemaker and saddler – just as John Carpenter was the village carpenter.

34 What White thought of as a similar case appeared from his historical studies of the parish. As its areas of woodland had been reduced over the centuries, the level of the Selborne stream had dropped, he believed. See Ch. 8, and Ch. 9, sect iii, below.

35 The so-called Gibraltar Correspondence; PF and Bibliography, 1985. See also my Preface, above.

Chapter 4: The Disappearing 'Swallows'

1 James Boswell, *Life of Samuel Johnson* (1768), Aetat 59.

2 TP xii. The 'Swedish author' seems to have been A. M. Berger, whose *Calendar of Flora for 1755* was included by Benjamin Stillingfleet in his *Miscellaneous Tracts,* 2nd edn (1762).

3 DB ix, and Sir Thomas Browne, *Miscellany Tracts* (1683), Tract V. ('North-*west* wind' is a slip of the pen, clearly.)

4 Francis Willughby, *Ornithology of Francis Willughby*, ed. and trans. John Ray (1678), Bk. II, p. 212. Sometimes referred to as 'Ray's Willughby', the *Ornithology* had been considerably augmented by Ray after Willughby's early death in 1672.

5 NJ Nov. 1789; see also Nov. 1782, Nov. 1777 and DB lv.

6 NJ 1780, Oct. 13 and 14.

7 DB lv.

8 NJ 1781, Apr. 6.

9 NJ 1793, Apr. 6.

10 E.g. NJ 1788, March 23. A similar search, with the same negative result, had earlier been described by Peter Collinson: 'Concerning the Migration of Swallows', *Philosophical Transactions* (1760), vol. 51 p. 459.

11 *Physico-Theology,* re-edited (1721), VII : iii, footnote.

12 'The Retreat of Birds of Passage', *Gentleman's Magazine* (1784), vol. xviii, p. 447. See Appendix, 2, above: on internal evidence, it seems very likely that White knew this paper. Cf. Peter Collinson's paper (1760), and a similarly sceptical letter to White himself from his friend William Sheffield, a fellow of Worcester College, Oxford; RH vol. I, p. 186.

13 TP x. A number of hibernation stories were summarised by Pennant, in *British Zoology*, 4th edn (1776), vol. II. Many others – including a great deal of 'under water' testimony – were provided by Daines Barrington in the paper, 'On the Torpidity of the Swallow Tribe', reprinted in his *Miscellanies* (1781), the work referred to by White at the beginning of DB li. For this important piece of Gilbert White's background, see Appendix, 3, above.

14 NJ 1778, Apr. 26. Derham, an *un*critical disciple of John Ray, is referred to on numerous occasions. He had edited new printings of Ray's *Wisdom of God in the Creation* and *Three Physico-theological Discourses,* and his own work, *Physico-Theology* (1713), itself went through several editions during the eighteenth century.

15 DB xxxvi.

16 DB xiii. White assumed Timothy to be a male, but it seems from the carapace, which can still be seen in the Natural History Museum, that 'he' may have been a female.

17 NJ 1787, Dec. 6.

18 Confusingly for the present-day reader, the binomial given by White for the swallow in *Selborne* is *Hirundo domestica*, and the house martin, *Hirundo rustica.* Today the swallow (or 'barn swallow') is *Hirundo*

rustica, and the house martin *Hirundo* (or *Delichon*) *urbica*.

19 It is now recognised that one or two bird species – though none which appears in Europe – can and do survive for long periods in a state of deep hypothermia and thus also torpidity.

20 DB xxxvi, and NJ 1777, March 26.

21 NJ 1774, May 6 and added notes; and June 3. In 1775, White had house martins which not only arrived late but arrived after the swifts: this militated on the side of migration. NJ 1775, May 3 and 13.

22 E.g. NJ 1788, Nov. 5.

23 RH II, p. 120. As we learn from the *Journal*, in the cold spring of 1784 the tortoise did not emerge until April 23, where in a warmer year he (or she) might have been out by the beginning of that month.

24 The late arrivals of 1774 are touched on in TP xl and DB xxii, but with no explicit references to the weather data, for example.

25 DB lv and li.

26 DB ix, and see also DB i. If Gilbert had had access to him, he would have received information of the same sort from William Bartram, the American traveller and naturalist: Bartram's *Travels Through Carolina, Georgia and Florida* was not published until 1791, but the work the book represented had been carried out in the 1770s. See Francis Harper, *The Travels of William Bartram* (New Haven, 1958), pp. 178–180.

27 DB xvii.

28 DB ix. White is responding here to the arguments Barrington had summarised in his paper, 'On the Periodical Appearing and Disappearing of Certain Birds, at Different Times of the Year', *Philosophical Transactions* (1772), vol. 62, p. 265.

29 NJ 1790, Apr. 11.

30 DB xvii, and NJ 1790, Apr. 11. Again, the *Journal* phrasing is more explicit.

31 See, for example, DB xxi.

32 DB xxi. As White notes, the differences had led J. A. Scopoli to suppose 'that this *species* might constitute a *genus per se*'.

33 TP xxvi.

34 See, for example, DB viii.

35 DB x. The passage from 'Ekmarck the Swede' translates as: 'the birds of the wading order, as if by agreement, fly away, nor can we find one residing here; for just as in the summer time the lack of earthworms and the hardness of the ground prevents their staying in hot countries, so, for the same reason, they cannot stay in cold winter climates'.

36 DB viii. Edward Jenner (1749–1823), was considering the same idea in the 1790s, although his thoughts on this topic were published only after his death; Bibliography, 1824.

37 TP xiii. Today, increasing numbers of blackcaps, though they move south in October, and are normally thought of as summer birds in Britain, now get no further than even the Scottish Borders in the course of their autumn migration.

38 TP xiii, and DB viii.

39 NJ 1775, May 3 and May 13. Lyndon was the home of White's brother in law and sister, Thomas and Mary Barker.

40 NJ 1774, May 6. At an early stage he had high hopes of Pennant, as someone with whom he could exchange information on bird movements in this way: see TP xiii. The European material on the activities of swallows and martins added to various of the *Journal* entries for March, April and May of 1787 was sent him by Dr Richard Chandler (1738–1810), during a lengthy period he spent at Rolle in Switzerland. See also the letter from the same source quoted NJ 1789, May 24.

41 'Of the Migration of British Birds', in Thomas Pennant, *British Zoology* (1776), vol. Il, Appendix, vi. White had added to and considerably improved on Pennant's original draught, in this appendix. (Pennant had more faith in Benjamin Stillingfleet than had White himself). Much of White's philosophy as a student of migration is explicit here.

42 See, for example, NJ 1770, flyleaf, where he quotes Linnaeus to the effect that observations in 'Southernmost Spain' will be important, and DB ix for the Bosphorus as a major crossing point.

43 He knew of course that birds could be marked; see DB ix. But in this connection, see also Ch. 6, note 40, below.

44 Eric Tomlin and Robert S. Steiner, *Bird Migration* (1978), Ch. 6.

45 DB viii.

46 TP xxxiii. That of John White was largely dependable information: with many of our summer migrants, the males precede the females and pairing takes place only after the males have established territories.

47 DB viii.

48 Cf. Eric Tomlin and Robert S. Steiner, *Bird Migration* (1978), Ch. 5, and Jean Dorst, The Migration of Birds (1962), Chs 2 and 8.

49 TP xli.

50 See, for example, TP xxxviii.

51 NJ 1793, Apr. 6 and 9.

52 In the *Journal* for 1769, whatever he was saying in letters to Pennant, he is quite positive concerning the ring ouzels: 'The ring-ouzels appear again on Noar-hill in their return to the northward: they make but a few days stay in their spring visit; but rest with us near a fortnight as they go to the Southward at Michaelmas.' NJ 1769, Apr. 13. For nightingales and swifts recorded on passage, see NJ 1778, May 16 and 1779, May 5 respectively.

53 NJ 1789, Sept 29.

54 The swallow is generally 'she', in the writings of Gilbert White.

55 NJ 1791, Sept. 14.

Chapter 5: Behaviour; Birds and Other Animals

1 TP xxx.

2 In his *Natural History of Northampton-shire* (1712), John Morton was

somewhat more generous. Although he too gives only a brief section to birds, Morton refers to two dozen species, and tells his readers where in his county they have been seen.

3 These works were often financed partly or largely by subscription, and engravings of country mansions might feature prominently among their illustrations.

4 William Borlase, *The Natural History of Cornwall* (1758, reprint 1970), p. 290.

5 *Ibid.*, pp. 289–90.

6 We have White's private summing up of Robert Plot: he 'is too credulous, some times trifling, some times superstitious; and at all times ready to make a needless display, and ostentation of erudition'. RH vol. I, p. 298.

7 DB vii.

8 NJ 1775, March 18.

9 NJ 1775, March 10.

10 NJ 1778, Sept. 4.

11 DB xxxix.

12 NJ 1780, July 28, and DB xxxix. Regarding the restricted numbers of swifts nesting at Selborne; W. H. Hudson found only 'eight or nine' pairs there in 1901.

13 DB iii. Which is not to imply, of course, that when there is no incubation there is no music: see DB iii and v.

14 DB xi. See also *The Natural History of Selborne*, ed. James Fisher (1941), notes 111 and 145. Fisher's notes, to this and to his later edition of *Selborne* (1947), are greatly to be recommended, as is his index and glossary of *Selborne* vertebrates. Incidentally, the term 'dispersion', used in this ornithological, or zoological, sense, was coined by White.

15 DB xliii. Cf. Edward A. Armstrong, *A Study of Bird Song* (1963), Ch. i and vii.

16 DB xviii and NJ 1776, Sept.18. The 'over-head calling' at night in summer was *eventually* identified as that of the stone curlews; see DB lix; and NJ 1788, March 22, and NJ 1790, March 20.

17 NJ 1777, Aug. 1.

18 NJ 1776, July 23.

19 DB vi. White shows his usual insight here, although today it is known that characteristic calls and postures also play an important part in this sexual distinguishing.

20 DB xxxii.

21 DB xlii and TP xxxix. It had not been established in his time that the 'bleating' sound was produced as air passed at speed through the snipe's extended tail feathers.

22 DB xliii.

23 DB lvii.

24 DB xxxii.

25 DB xx.
26 TP xi and vi.
27 NJ 1775, Apr. 1, and 1777, Apr. 20.
28 DB xlviii.
29 NJ 1789, Nov. 30.
30 DB iv. This letter is dated Feb. 19, 1770, in *Selborne*. It includes additions made after this date, but from White's opening remarks we can take it that he owes nothing here to Edward Jenner, or the latter's important paper on the cuckoo (1788). Jenner, a physiologist and medical researcher, was only an occasional ornithologist – unfortunately for ornithology.
31 DB iv.
32 In July 1771, he was telling Pennant that if he were to attempt publication himself, an idea he still viewed with great diffidence, it might be of a 'journal for one year'. See also Ch. 7, sect. i, below.
33 TP xvi, xxi and xxxiii.
34 NJ 1774, Jan. 14 and NJ 1793, Feb. 24.
35 NJ 1789, Aug. 9. Cf. RH vol. II, pp. 205–209.
36 He had heard the bird's churring call while he also watched a nightjar at rest, for 'many a half hour'; but how could he be quite sure the bird he was watching was the bird he could hear? He covers the point as he adds, 'it sat with it's under mandible quivering'. TP xxii.
37 NJ 1773, July 31; TP xxxvii; and NJ 1792, Aug. 27. The hobby falcon, which takes large insects, will behave in the same way.
38 NJ 1770, March 27. Oddly, he failed to register the faint, high-pitched song of the spotted flycatcher: see TP xl.
39 Referring to Stillingfleet's *Miscellaneous Tracts*, he says: 'Mr. *Stillingfleet* makes it a question whether the *blackcap (motacilla atricapilla)* be a bird of passage or not: I think there is no doubt of it'. TP x. (The classification 'Motacilla' seems strange today; the blackcap is now a 'Sylvia'; but the *observation* is, as usual, sound.)
40 TP x.
41 For the bat, *Nyctalus noctula*, see TP xxvi and TP xxxvi. For the lesser whitethroat, DB lvii.
42 TP xii, xiii. The animal became *Micromys minutus*.
43 See TP xvi and xix.
44 DB vi. Cf. the extensive discussion of taxonomic matters in the letters to John White, 1769–72. See, for example, my Preface, above.
45 NJ 1775, Feb. 2. For the condemnation of the rook among many other species by Act of Parliament, see Ch. 8, note 1, below.
46 DB v.
47 Stephen Hales, 'Vegetable Staticks', *Statical Essays* (1733), Ch. VI, Experiment cxiv. For a lengthy extract from the same grisly report, see Appendix, 1, above.
48 NJ 1776, July 14.
49 NJ 1775, July 1, and 1776, Dec. 3. On the question of earthworms'

remaining anchored, White seems to disagree with Charles Darwin. See *The Formation of Vegetable Mould Through the Action of Worms* (1881), Ch. 1.

50 DB xlix.

51 TP xvii and xviii; also DB xxxi. Richard Jefferies mentions the same alleged remedy uncritically.

52 NJ 1790, Oct. 8.

53 DB xlii. The present-day bird watcher refers to the 'jizz' of a particular species. See James Fisher/Jim Flegg, *Watching Birds* (1974), Ch. 3.

54 DB xv.

55 DB xvi; and NJ 1776, Apr. 26. *Hirundo apis* was the swift (now *Apus apus*). In the Introduction to his edition of *Selborne* (1929), E. M. Nicholson provides a particularly valuable review of White's bird records.

56 DB xv.

57 DB xvi.

58 *A Herd of Red Deer* (1936), Ch. II.

59 NJ 1775, Nov. 3 and TP xl. Today the British list hovers around 470 bird species.

60 NJ 1793, Jan. 20. White started his correspondence with Marsham only in 1790: the two turned out to be kindred spirits, and the Selborne naturalist greatly regretted that their friendship had not begun 'forty years ago'. See RH vol. II, Chs x and xi. For the black-winged stilts, see TP xi, and NJ 1779, Apr. 28.

61 NJ 1792, June 5.

62 See, for example, NJ 1784, May 28, and NJ 1786, May 25.

63 DB xvi.

64 NJ 1770, Feb. 13. See also NJ 1773, Dec. 15.

65 In lists sent to Pennant (TP xvi) and Barrington (DB i) in the late 1760s, White seems to include the corncrake as a breeding bird in his neighbourhood; but in September 1789, he refers to the species in the *Journal* as 'rare in this district' and seen only in autumn. For the honey buzzard, see, for example, TP xliii and NJ 1787, Sept. 12.

66 See, for example, NJ 1780, Nov. 23.

67 E.g. DB i. In the eighteenth century, as is still the case, nightingales, though regular summer visitors not only to Hampshire but to most of England, did not spread over the whole of Great Britain. DB ix.

68 E.g. TP xliv and NJ 1770, Nov. 24. Strangely to us, White was sure that no stockdoves remained in his district during the breeding season. The stockdove (*Columba oenas*) is also called the wild or wood pigeon, in his writings; what we know as the woodpigeon (*Columba palumbus*) is usually called the ringdove.

69 TP vi.

70 NJ 1776, Sept. 6.

71 TP xl.

72 TP xli.

73 TP xli.

74 NJ 1784, Aug. 25. On other occasions a dog ate gooseberries or plums, and some partridges which bred on White's property ate raspberries. See NJ 1785, Sept. 1 and 1788, Sept. 23.
75 NJ 1781, Aug. 19 and 1791, Sept. 3.
76 NJ 1791, Sept. 3. He can also show that this species does not depend exclusively on an insect diet; TP xliii.
77 See NJ 1785, July 13; and 1781, Aug. 24 and 31.
78 NJ 1776, Sept. 18.
79 DB x.
80 NJ 1780, March 14.
81 See Ch. 6, note 15, below.
82 NJ 1791, March 30 and Aug. 7.
83 See, for example, TP xxxix.
84 NJ 1793, May 9. Other notable items in the series are dated 1774, April 14; 1780, May 12 and 18; and 1784, May 6.
85 D.B xxxiii.
86 TP xiv, and DB v. However, my father, a very 'intelligent' countryman, agreed with the warreners quoted by White, that 'rabbits are never in such good case as in a gentle frost'.
87 This hand rearing is by no means to be confused with professional trapping, then.
88 TP xi, and NJ 1788, May 11. Today we would say, 'only mice and field voles'.
89 TP xi.
90 DB liv. Cf. his notes on the tame bat; TP xi.
91 The 'nearby menagerie' was that of John Hunter, at Earl's Court. Hunter was a surgeon and comparative anatomist; the flamboyant behaviour for which he was well known would perhaps not have deterred White, but he was a ready dissector of living as well as dead animals.
92 TP xxviii.
93 See Ch. 6, note 15, below.

Chapter 6: Instinct and Initiative

1 NJ 1775, March 18.
2 David Lack, *The Life of the Robin* (1943), Ch. vi.
3 NJ 1780, May 6.
4 See, for example, W. H. Thorpe, *The Origins and Rise of Ethology* (1979).
5 The confusing term 'Behaviourism' should perhaps be mentioned here. It was used by J. B. Watson and others early in the twentieth century to stand for their laboratory work. We can approve of Watson's insistence on observables and measureables, and thus 'behaviour', but he restricted himself to the reactive behaviour he associated with reflexes, and constructed a hypothetical (behaving) animal from material of this kind. Inevitably, this animal had a machine-like character – and drew suspicion

on the idea that life and conversation could be examined scientifically.

6 See, for example, Konrad Lorenz, *King Solomon's Ring* (1952), and Nikolaus Tinbergen, *The Study of Instinct*, revised edn (1969). The introduction to the 1964 reprint of *King Solomon's Ring*, by W. H. Thorpe, remains itself an admirable short introduction to ethology.

7 DB lvi. See David Lack, *The Life of the Robin* (1943), Ch. xv.

8 DB xiv.

9 DB xiv and xxxi. The tests conducted by Douglas Spalding on newborn chicks and piglets in the mid-nineteenth century were the first systematic work on the same topic. See Bibliography.

10 DB xxxi. This is almost a quotation from John Ray: *The Wisdom of God in the Creation* (1691), p. 93.

11 NJ 1793, Feb. 24.

12 In the British psychological tradition often reflected in White's phrasing and concepts, all behaviour has something of conation – and emotion – about it. See Ch 11, sect. 2 below.

13 DB v. In connection with the same behaviour, Lorenz refers to 'imaginary' eggs.

14 DB xvi. See also DB xviii.

15 'Experiments and Observations on the Singing of Birds', *Philosophical Transactions* (1774), vol. 63, p. 249; re-printed in Pennant's *British Zoology,* 4th edition (1776), vol. II, Appendix v.

16 TP xii.

17 DB xlvi. When he nipped out the buds of a summer-flowering rose bush to find out whether it would flower in the autumn, no later flowering occurred; NJ 1784, June 5.

18 See W. H. Thorpe, *Learning and Instinct in Animals* (1956), and Edward A. Armstrong, *A Study of Bird Song* (1963), Ch. iii.

19 DB xiv.

20 NJ 1791, Aug. 21.

21 DB lvi. The notes on these chaffinches and wrens in the *Journal* were made at South Lambeth; NJ 1787, May 27.

22 TP xxi and xxii. He means in south-west Sussex and south-east Hants; and he could have added 'and there are few rocky clefts'.

23 DB xiv.

24 DB xxiv.

25 DB xxiv.

26 DB xxxiv and NJ 1790, April 18. Size for size, he says, the cat is normally 'carnivorous and predaceous', a 'bloody grimalkin'.

27 Konrad Lorenz's goslings – or perhaps those reared by his teacher, Oscar Heinroth – inevitably come to mind, the goslings which imprinted their human 'parent'.

28 DB xxxiv. The cat with the newborn squirrels, similarly, 'supposes the squirrels to be her own young': NJ 1790, April 19. Cf DB vii, the 'titlark' which was 'the dupe of a dam'.

29 See, for example, Nikolaus Tinbergen, *The Study of Instinct*, revised edn

(1969), Chs 3 and 5.

30 DB xi.

31 DB xi.

32 DB lii and xxiv.

33 See, for example, DB xiv.

34 He goes to the Greek, *στοργη* (storge) for an adequate term for felt maternal or parental attachment. See, for example, DB xviii and xiv.

35 E.g. DB xviii.

36 NJ 1781, July 7: see also 1777, July 18, and 1781, July 22. Lorenz's jackdaws showed a similar concern for their young. See *King Solomon's Ring* (1952), ch 11.

37 TP xxxiii.

38 NJ 1778, May 21; and TP xl.

39 See, for example, NJ 1787, Sept. 5.

40 NJ 1775. June 3, and 1792, May 22. See also NJ 1775, Apr. 27. It has been pointed out to me that Daines Barrington suggested to White that he should mark some of his 'regularly returning' birds. White did not act on this suggestion; but during the 1780s, Edward Jenner marked some swifts, with incisions on the beak, and proved that some of these birds did return to the same nesting sites. This work too was published only in 1824, in Jenner's posthumous paper.

41 DB xxviii.

42 See the uninhibited asides in PF pp. 320, 323 and 325.

43 *Realities of Bird Life* (1927), Ch. xv.

44 If other and later experimentalists had found that pressure *might not* be inversely proportional to volume, in a given quantity of gas at a given temperature, there would be no received 'Boyle's law'.

Chapter 7: Behaviour; Plants and Insects

1 TP xvii and DB xxxv. At least two expurgated versions of *Selborne* were produced in the nineteenth century; see, for example, J. G. Wood's edition (1860); and more extraordinarily, one of them has re-emerged in recent years. The unexpurgated text includes, along with the venereous earthworms, the copulating swifts, DB xxi; the fact that the male frog has 'no penis intrans', TP xvii; a note on the great bat's 'parts of generation', TP xxxvi; and two paragraphs on castration and related topics, DB xxxii.

2 NJ 1770, Aug. 9.

3 NJ 1787, July 14.

4 NJ 1791, Aug. 3. In 1788, Hale had had a 'prodigious' crop of hops in the same place, 'under the s. corner of the hanger'; NJ 1788, Sept. 2.

5 It is widely accepted in Britain, but not on the Continent, it seems. See E. J. Jeffery, *Brewing, Theory and Practice* (1956), and WJ Ch. Eight.

6 NJ 1774, April 30.

7 NJ 1788, April 12.

8 The present-day gardener, if he is not using F1 hybrids, picks off the male blossoms of frame cucumbers, preventing pollination and thus avoiding 'big-endedness'; but it is possible that the frame varieties of White's time were closer to outdoor cucumbers than are the present-day frame varieties.

9 NJ 1790, July 24.

10 NJ 1778, Nov. 12.

11 TP xvii. It was 'obscure', as were the histories of truffles and that 'strange jelly-like substance', common nostoc., for example, NJ 1783, Sept. 4, and 1791, Nov. 29.

12 GC 1767, Aug. 17. In the *Naturalist's Journal*, begun in 1768, he names and classifies insects according to Linnaeus and plants according to the 'sexual system', but keeps largely to Ray's nomenclature for birds and other animals.

13 See WJ Ch. Eight. Johnson was original, and is particularly authoritative, on White as botanist.

14 Benjamin Stillingfleet, *Miscellaneous Tracts Relating to Natural History, Husbandry and Physick*, 2nd edn (1762), Preface.

15 In a case White refers to explicitly, members of one natural genus, the numerous 'thistle family', could be *artificially separated*, where ratios of sexual parts were the criterion. A letter to John White written in 1772; PF p. 496.

16 DB xl and PF pp 495–96.

17 Ibid., Stillingfleet's introduction to his own 'Calendar of Flora'.

18 See, for example, PN 1788, July 12. The passage from Virgil has been rendered as:

> Observe again, when the walnut clothes herself in the woods
> With richest bloom and bends to earth her scented branches –
> If her fruit is plentiful, a plentiful corn-crop follows,
> And great will be the threshing in a season of great heat.
> *The Georgics,* trans. C. Day Lewis, (1943), Bk I, lines 187–190.

19 DB xli.

20 DB xli. The significance of the species name would not have been lost on White – the plant also favours pine woods.

21 DB xli. Also NJ 1779, Apr. 21, and 1780, May 8.

22 Linnaeus made little advance towards a 'genealogical' classification, no doubt because of his adherence to fixed species; see Ch. 9, sect. 4, below. By comparison, his introduction of 'binomials' – each described creature being given two names, one generic and one specific – has been universally accepted; although, since his time, numerous of his own binomials have been altered and improved on, and a great many new ones have been added.

23 NJ 1789, July 19 and 21.

24 NJ 1791, Jan. 1, and 1792, Nov. 17.

25 TP xxvi. The animal is likely to have been not the 'water rat' or water

vole, *Arvicola amphibius amphibius*, which does not store food, but that active incomer, the brown or Norway rat, *Rattus norvegicus*.

26 DB xix.

27 W. N. Turrill, *British Plant Life* (1962), Ch. 11.

28 NJ 1782, Sept. 2.

29 NJ 1784, Nov. 14.

30 NJ 1783, Aug. 24.

31 DB lxi.

32 NJ 1775, Sept. 27.

33 NJ 1791, July 24. A larger relative of the cuckoo pint, it is now named *Dracunculus vulgaris*.

34 GC 1766, March 26.

35 See, for example, NJ 1776, March 22.

36 DB xli. Apart from one or two pejorative remarks, White makes virtually no references to the medicinal uses of specific plants.

37 TP i.

38 TP ix. See also TP ii, and Ch. 9, sect. 3, below.

39 TP i. NJ 1779, June 1, and 1784, Apr. 20. This was one of his points of discussion with Robert Marsham, in the exchanges which took place from 1790 to 1793. See RH vol. II.

40 NJ 1781, July 18 and NJ 1783, July 3.

41 NJ 1791, June 21.

42 DB xli. The second half of this paper appeared in the first edition of *Selborne* (1789), but as an 'addition' placed before the appendixes.

43 DB xlvi, xlvii and xlviii.

44 See Bibliography. In the late seventeenth century, both Ray and Swammerdam left precisely observed notes on insect behaviour: see, for example, C. E. Raven, *John Ray, Naturalist* (1942), Ch. xv.

45 TP xxii, and, for example, PF p. 321.

46 Quoted by David Allen, *The Naturalist in Britain* (1979), Ch. Seven. As Allen says, the habits of these 'entomologists' were themselves examples of ritual.

47 DB v.

48 PF p. 316. Cf. the recipe for an 'antiseptic substance for the preservation of birds', also noted with John White in mind, on the flyleaf of NJ 1770.

49 NJ 1776, Apr. 4.

50 NJ 1784, Sept. 5.

51 NJ 1770, July 21.

52 NJ 1772, July 11.

53 NJ 1776, June 26.

54 DB liii. The last of the consignments of specimens from John White had arrived in 1772. The 'coccus' is a beetle; White gives the Linnean binomial *Vitis viniferae*, more recently it has been *Pulvinaria vitis*.

55 WJ Ch. Seven. For some perceptive work on a Spanish variety of

lacewing, *Neuroptera bipennis*, found by White among John White's specimens, see Paul G. M. Foster, *Gilbert White and his Records* (1988), Ch. Ten.

56 DB xxxviii.

57 DB xxxviii.

58 NJ 1780, Sept. 17. This presumably is the source of the more recent references to White's 'playing the trumpet' to animals.

59 DB lxvi and NJ 1790, March 31.

60 DB xlvi and xxxviii.

61 A further stage can be added to this story, I find. In an unfinished paper, White looks at evidence concerning *'tame fishes'*: the 'best modern Ichthyologists' assure us, he says, that fishes have no organs of hearing, 'But then if fishes do not hear, some will say how do tame fishes in stews and canals come to be fed at the sound of a whistle?' Sensitivity to vibration is the key, he believes; 'hearing and feeling' are kindred senses. *The Natural History and Antiquities of Selborne*, ed. Thomas Bell (1877), vol. II. By the time *Selborne* was published, the views of the 'best ichthyologists' had changed; but see also DB liv.

62 NJ 1780, Aug. 2. In present day Britain, the swallowtail occurs only in parts of Norfolk, but areas of marshland habitat were more common in the eighteenth century than they now are.

63 DB xlvi. See also GC 1761, May 20. Thirty years later, in 1791, May 29, White was to make the following entry in the *Journal*: 'The race of field-crickets which burrowed in the short Lythe, & used to make such an agreeable, shrilling noise the summer long, seems to be extinct'. He blames himself, he thinks the village boys have been using his method of pulling the crickets out of their holes; but he could have referred also to the long sequence of wet summers from 1764 to 1774.

64 In another age, White might have added writing for children to his other activities. In 1784, in reply to some verses from John Mulso's daughter, he did write the *Letter from Timothy the Tortoise to Miss Hecky Mulso*, giving a tortoise's eye view of life at The Wakes. RH vol. I, p. 128.

65 NJ 1772, Aug. 22: compare with NJ 1789, May 10.

66 NJ 1788, June 14; and 1785, June 19. The moth seems to have been *Tortrix viridana*, which still occasionally occurs in great numbers in southern England.

67 TP xxxiv.

68 PF p. 321, and TP xxxiv. Moses Harris's practical pocket book, *The English Lepidoptera*, appeared only in 1775, and was of course *confined to* these genera.

69 PF p. 323, and TP xxxiv. There has been confusion as to who White meant by the 'Geoffroy' he mentions on several occasions, some writers settling for the earlier and less sophisticated S. F. Geoffroy (1672–1731). The mistake has been due in part to the inadequate reference provided in the first edition of *Selborne,* and carried forward in all subsequent editions; after White's remarks on the *Stratiomys chamaeleon*, we have, 'see Geof-

froy, t. 17, f. 4'. The reference is to a *plate*, and takes us – eventually – to a diagram of the chameleon fly, and the 'star-like tail', or breathing tube, of the larval form, in the *Histoire abrégée des insectes* (Paris, 1762), by the physician and entomologist Etienne Louis Geoffroy (1725–1810). This innovative and particularly well illustrated work is one recommended also in the 'Gibraltar letters'. See, for example, Preface, p xi, above.

70 DB xlviii and xlvii.

71 DB xlviii. The more recent understanding of the house cricket is that its essential adaptation was to the sub-tropical climate of north Africa and western Asia: it arrived in northern Europe, then, and survived in human habitations. White's attention to this 'domestic insect' is, nonetheless, thoughtful and perceptive.

72 NJ 1778, July 11 and Dec. 24.

73 NJ 1792, Aug. 18

74 See TP xxxiv. The minute 'harvest bug' is the larva of *Trombicula autumnalis*, it seems, and the horse bot or gad-fly, the 'oestrus, known in these parts to every plough-boy', is *Gastrophilus equi*.

75 NJ 1775, Sept. 9.

76 NJ 1774, May 28.

77 NJ 1777, Feb. 9, and 1791, Oct. 9. Flies exude a sticky substance; they do not 'depend on atmospheric pressure'. White may have been misled by Swammerdam here, a copy of whose *Book of Nature and History of Insects* (1758) he owned.

78 NJ 1783, June 7

79 DB lxiv, and NJ 1783, May 17.

80 NJ 1774, Sept. 12.

81 NJ 1771, Jan. 3.

82 DB liii.

83 DB xxiii. These aerial spiders are 'apterous insects' in that they fly in spite of 'not being fitted with wings': White is referring to an interesting fact here – he is not somehow associating these spiders with springtails or lice.

Chapter 8: The Useful Naturalist

1 Dozens of blackbirds might be shot on occasion; see, for example, PN 1781, Aug. 8, and 1783, Aug. 1. As some mitigation, 'troops' of birds could descend on a garden in Hampshire in White's time, and by current standards his action was, if anything, restrained. The systematic 'thinning' of birds and animals which seemed to compete with man was still legally sanctioned and even enjoined. Among the dozens of proscribed species, with all the owls and corvids and most of the hawks, were 'woodpeckers', kingfishers and even goldfinches. See, for example, Colin R. Tubbs, *The Buzzard* (1974), Ch. 2.

2 RH vol. i, p. 328. Cf. Benjamin Stillingfleet, 'Observations on Grasses', in *Miscellaneous Tracts,* 2nd edn (1762).

3 DB xvii.

4 NJ 1776, Aug. 30.

5 NJ 1772, May 16.

6 NJ 1781, Nov. 7.

7 *Minutes of Agriculture, Made on a Farm of Three Hundred Acres of Various Soils* (1787); Digest, Pt One.

8 See, for example, NJ 1790, Aug. 20, Aug. and Aug. 31. In Oxfordshire, though, corn ricks were in limited use even in Robert Plot's time.

9 GC 1765, April 13.

10 Potatoes were by no means universally accepted in England in the mid-eighteenth century, but several varieties were grown in the parish, for human and animal consumption, as a result of White's efforts. In the autumn of 1787, with heavy crops of both beech nuts and potatoes, he notes, 'Between mast and potatoes, poor men killed very large hogs at little expense. Tom Berriman's hog weighed 16 scores, yet ate only *seven* bush. of barley-meal; whereas without the help above mentioned he would have required 20 bush[s].' (A 'score' in this context is twenty pounds weight.) Letter reprinted in *The Natural History and Antiquities of Selborne*, ed Thomas Bell (1877), vol. II. See also NJ 1787, Oct. 3 and DB xxxvii.

11 NJ 1789, Dec. 6.

12 NJ 1771, Feb. 28.

13 NJ 1771, March 2.

14 *Naturalist's Journal*. 'Four quarters on the acre' meant 32 bushels from an acre – which would have been a goodish crop until well into the twentieth century.

15 Sainfoin was first cultivated widely in Britain in the late seventeenth century. It was among the leguminous, and as was later to be established, nitrogenous, field crops which could be grown beneficially on land which would traditionally have been left idle for a season after corn-growing.

16 NJ 1792, Sept. 8.

17 An awareness of these leading lights can be taken for granted, although Robert Bakewell became famous rather later than the others. (The touring agriculturist Arthur Young (1741–1820) is referred to explicitly in the correspondence with Robert Marsham.) For the 'agricultural revolution', see, for example, T. S. Ashton, *An Economic History of England; The Eighteenth Century*, revised edn (1972).

18 See *The Invitation to Selborne*, included in the selection of White's poetry added to the 2nd edition (1813), and many subsequent editions, of the *Natural History*.

19 NJ 1788, June 28. The *seed catalogues* of the period could be astonishingly ambitious: 'in 1780, for example, it was possible to buy as many as 320 different kinds of gooseberry . . . and in 1777 Richard Weston's seed catalogue offered 208 anemones, 575 hyacinths, over 800 tulips and no less than 1100 ranunculi'. Keith Thomas, *Man and the Natural World, 1500–1800* (1983), Ch. v : iv.

20 TP xxii.

21 In 1752, we know, at the beginning of his early gardening at Selborne, he paid two visits to Stowe, in Buckinghamshire: on these occasions, at least, he saw the work of Bridgeman, Kent and Brown at first hand.

22 In their attempts to encourage further tree-growth on the Down, they met with the problem which has so often bedevilled tree-planting, that of grazing animals. As White records, 'As fast as any of these seeds have sprouted, they have constantly been brouzed off, & bitten down by the sheep'. NJ 1789, Oct. 22.

23 GC 1792, Dec. 18. In later years, though, the flooding re-occurred.

24 GC 1792, Dec. 18. White was no doubt encouraged in these activities by the example of his grandfather, who became vicar of Selborne in 1681 (he was a member of Magdalen College). This earlier White renovated the Selborne vicarage and left money for road repairs and repairs to the church. He left £100, in addition, 'to be laid out on lands; the yearly rents whereof shall be employed in teaching the poor children of *Selbourn* parish to read and write, and say their prayers and catechism, and to sew and knit'. AN vi.

25 A view of Selborne as it seemed to the visitor early in the nineteenth century is provided by William Cobbett. He had an unfortunate experience with a muddy lane while crossing the south-western end of the parish in 1822, but when he visited the village for the first time, in August 1823, he found it as interesting, and attractive, as White's description had led him to expect. See *Rural Rides* (1826), 1823, Aug. 7.

26 NJ 1774, Apr. 1, and 1775, Apr. 15. Also AN ii.

27 TP vi and DB lix. For the present-day Woolmer Forest, see Ch. 9, note 73.

28 Not, the water level had been higher, and so there had been more trees; but, there had been more trees, and because of their 'operation as an alembic', there had been more water in the stream.

29 AN xxv.

30 AN viii.

31 See the Advertisement, or Foreword, to *Selborne*, and DB lxvi. Robert Plot and John Morton included 'some account of the Antiquities' within their county 'natural histories', although in his *Oxfordshire* Plot confines himself mainly to pre-medieval remains. William Borlase's study, *The Antiquities of Cornwall* (1769) was published separately from, but was part of the same overall project as, his natural history.

32 See the Advertisement to *Selborne*.

33 Walter S. Scott, *White of Selborne and his Times* (1946), Ch. 1.

34 See AN x and xxvi.

35 H. J. Massingham over-emphasises the medieval character of eighteenth century Selborne, but the 'tithe award map' he published, and his notes on this map, give us valuable information on the village and parish of White's day. *The Writings of Gilbert White of Selborne*, ed. H. J. Massingham (1938), Appendix.

36 TP ix and NJ 1784, Apr. 4. Also RH vol. II, Ch. v.

37 The population of Britain gradually increased during the eighteenth century, approximately doubling between 1700 and 1800, but the sharp upward movement in corn prices and living costs generally began only with the outbreak of war with France, in 1793.

38 DB xxxvii.

39 DB xxxvii. The forward-looking and 'moral' agriculturalist is admirably described by Tobias Smollett, in *Humphry Clinker* (1771), letter dated Oct. 11.

40 DB liv.

41 *The Invitation to Selborne*.

42 Cf. Alexander Pope, *Moral Essays* IV (1731), lines 47–78; and Virgil, *Georgics,* Bk. I, lines 43–256,, for example, in C. Day Lewis's translation (1943). The perverse and misguided agriculturalist – he lives to regret it – is also depicted by Smollett: see *Humphry Clinker* (1771), letter dated Sept. 30.

43 A sermon of White's, quoted in *The Natural History and Antiquities of Selborne*, ed Thomas Bell (1877), vol. II.

Chapter 9: Science, Meteorology and Geology

1 Good editions of *Selborne* include these population and mortality figures, and print them in their original place, as a footnote to TP v. See Sect. 2, below.

2 TP viii.

3 See, for example, TP i, or NJ 1781, Sept. 14.

4 GC 1765, Apr. 29. Several writers have noticed that in DB xlix there is an error in the calculation concerning the legs of the black-winged stilt; but the calculation, and not the error, was typical of Gilbert White.

5 The one exception to this in *Selborne* – the reference to the child who thinks the rooks are 'saying their prayers', in DB lix – highlights the character of the rest of the book.

6 A book 'composed of letters' was an eighteenth century convention, I have been reminded, but *Selborne* relates as much to the *Philosophical Transactions*, the annual collection of papers or 'letters' read before the Royal Society, as to this convention. The *Selborne* letters, and the assumed replies, are letters to and from collaborating naturalists.

7 One of the illustrations prepared for the first edition of *Selborne*, we notice, was rejected by White as 'too romantic'.

8 I regard as misleading, then, the description of White's research as 'pursued primarily . . . to show, experimentally, the bounty of Providence'. Paul G. M. Foster, *Gilbert White and his Records* (1988), Ch. One. Even implicit sermonising is largely absent from *Selborne*, which in this respect can be compared and contrasted with, say, Humphry Primatt, *The Duty of Mercy and Sin of Cruelty to Brute Animals* (1776 and 1992): for an extract from the latter, see Appendix, 4, above.

9 DB xxviii.

10 *Ibid.*

11 TP xviii. But as he says in the same paper, various 'intelligent persons' seemed to accept the story.

12 AN iii.

13 TP xxi.

14 For the extent of Thomas's support, see C. S. Emden, *Gilbert White in his Village* (1956), Ch xii.

15 David Elliston Allen, *The Naturalist in Britain* (1979), Ch Two.

16 British thinkers associated this notion of 'science' also with Newton, correctly or otherwise.

17 *An Essay Concerning Human Understanding* (1690), Bk IV, i, 2, and Bk II, xxv, 1–10. Cf. David Hume, *A Treatise of Human Nature* (1739), Bk I, iii, ii.

18 For a general statement of what I am calling empiricist scientific principles, see George Berkeley, *The Principles of Human Knowledge* (1710), sects ci–cxvi. Berkeley was a theologian in one of his roles, and perhaps an idiosyncratic one, but he expressed with especial clarity the idea that natural science – predictive 'natural philosophy' – could be logically independent of the discussion of non-observables. Hume, the third of the three great empiricists, takes this as axiomatic. As methodologists, he says, 'We may define a CAUSE to be "An object precedent and contiguous to another, and where all the objects resembling the former are plac'd in like relations of precedency and contingity to those objects that resemble the latter".' *A Treatise of Human Nature*, Bk I, iii, xiv. Cf. ibid. Bk I, iii, xv.

19 See Locke, *An Essay Concerning Human Understanding* (1690), Bk II, xxxii, 9, 14 and 15; and Bk IV, xvi, 6–7. Then, for example, Bk IV, xi, 4–8 and Bk IV, xiii, 2.

20 Berkeley develops the 'universal language' idea in the *Principles*, having approached it first in *A New Theory of Vision* (1709); see, for example, *New Theory,* sects cxliv and cxlvii.

21 John Locke, *An Essay Concerning Human Understanding* (1690), Bk IV, xix, 1–12. Cf. Hume, *A Treatise on Human Nature* (1739), Bk III, ii, vii. 'What suits our wishes is forwardly believed', Locke says. For some detailed early eighteenth-century studies of self-deceit, see Joseph Butler, *Fifteen Sermons* (1726), VII and X.

22 See, for example, Hume's famous *Essay on Miracles*, included in David Hume, *Essays Moral, political and Literary* (1963).

23 TP xxiv.

24 DB xliii and DB xlvi.

25 Specialist natural history periodicals were unknown in White's time, but see his own suggestion; RH vol. I, pp. 170, 171.

26 *The Wisdom of God in the Creation* (1691), p. 130.

27 DB x.

28 *The Wisdom of God in the Creation* (1691), pp. 125,26.

29 Despite the activities of his brothers and friends, Gilbert could remark in an early letter to Pennant, 'Mr. Skinner of C.C.C. and Mr. Sheffield of Worcester College have lately been with me for a fortnight, and are the only Naturalists that I have ever yet had the pleasure of seeing at my house'. RH vol. I, p. 171. Several of William Sheffield's letters to White have been preserved. Like Gilbert White's own letters to his brother John, they show us the technical competence assumed in some of White's private correspondence. See Paul G. M. Foster, 'William Sheffield Letters to Gilbert White', *Archives of Natural History*, vol. 12 (1985), pp. 1–21.

30 Stephen Hales, *Vegetable Staticks* (1733), Preface.

31 For this reason too, the notion that he was in some sense an 'opponent' of the Royal Society is quite misleading. See *The Writings of Gilbert White of Selborne*, ed. H. J. Massingham (1938), Introduction.

32 For various ballooning episodes – including some trials of model balloons on Selborne Down – see NJ 1784, Sept. 1, Oct. 16 and Oct. 21; also the newspaper cutting pasted into the *Journal* opposite the Oct. 10–16 entries.

33 That the relationship was cordial appears no doubt from White's spending more than four months in Rutlandshire with Thomas and Anne Barker early in 1760.

34 See Ulrich Trohler, 'To Improve the Evidence of Medicine', *History and Philosophy of the Life Sciences* (1988), vol. 10, supplement, p. 31. An important example of this 'early epidemiology' was William Black, *An Arithmetical and Medical Analysis of the Diseases and the Mortality of the Human Species* (1789). White himself refers familiarly to John Huxham's discussion of 'meteorology and morbid epidemics' (1749, see Bibliography), in DB lx.

35 There was no confusion in his own mind here. In private, he could describe some of the leading naturalists of his time as 'very barren' regarding behaviour. PF p. 323.

36 See TP xvi and xxxiii. In the same way, he implies that John White is a quite independent source in the matter of the vine coccus, for instance in DB liii.

37 F. Fraser Darling, *Wild Life in Britain* (1943), p. 27.

38 DB lxi.

39 NJ 1790; this page is out of sequence in the *Journal* as now bound, but seems to go with Oct. 3–9.

40 NJ 1784, Dec. ll.

41 NJ 1790, June 22. White's own taking of temperature readings could be somewhat erratic; he varied considerably in the number of readings he took during a day, from one to four. But the point here is that temperatures at Selborne even approaching these at South Lambeth would have been entirely exceptional.

42 The following, from the relevant *Journal* volumes, can be added to the annual rainfall figures given in *Selborne*, TP v.

1787	36.24
1788	22.50
1789	42.00
1790	32.27
1791	44.93
1792	48.56

43 NJ 1776, Jan. 14.

44 NJ 1776, Jan. 17–20, and DB lxii.

45 DB lxii.

46 See DB lxi.

47 DB lxiii. In this 'letter', White refers to the difference in elevation as 'two hundred feet or more'. By his own calculation it was three hundred feet, but the candid naturalist knew that others had made it little more than two hundred; see DB lx.

48 DB lxiii, and *Journal*.

49 DB lxiii. The winter of 1739–40, the 'killing winter', had become a legend for severity.

50 W. B. Turrill, *British Plant Life* (1962,) Ch. 11.

51 But he did not forget these strange events. An unseasonable white frost in June 1791 damaged garden plants at Selborne but left those at Newton untouched; NJ 1791, June 14 and 15.

52 NJ 1783, June 7.

53 NJ 1781, Oct. 14. Even with some wet weather during that winter, the rainfall for 1781 reached only 30.71 inches.

54 The 'lavants' – the occasional springs which break out on the sides of hills in White's region after a persistently wet winter season – are explained in this way.

55 DB xxix. He states the problem first in NJ 1772, Aug. 15. After this he touches on it or data relating to it in the following *Journal* entries, among others: 1775, May 27, July 14 and Oct. 6; 1776, May 25; 1778, Dec. 9; 1781, Oct. 14. This is not to say that his was a complete explanation, 'dew ponds' are still something of a mystery.

56 RH vol. I, p. 295.

57 NJ 1782, Nov. 14, and 1790, Aug. 7.

58 DB lxvi, or, for example, NJ 1783, July 10.

59 DB lxvi. See also NJ 1784, June 6.

60 NJ 1785, March 7.

61 See *Journal* notes.

62 DB lxv.

63 DB xxxviii and lx. By comparison, John Aubrey, could find 'no polysyllabical echoes' in Wiltshire.

64 RH vol. II, p. 159.

65 NJ 1771, March 23.

66 Letter published in *The Natural History and Antiquities of Selborne*, ed. Thomas Bell (1878), vol. II. By this date John had returned to England: the 'American' material would be found in Kalm's *Travels into North*

America (1770).

67 PF p. 315.

68 TP xl. Linnaeus was here making one of his various later suggestions for a more 'natural' botanical taxonomy.

69 There were 'theories of the world', but methodically established geological accounts had to wait for James Hutton (1726–97) and the 'father' of the discipline in England, William Smith (1769–1839). Smith's 'stratigraphical' map, the first modern example, was published only in 1815.

70 DB xvii.

71 Noar Hill, just on the Newton Valence side of the parish boundary, reaches almost 700 feet above sea level.

72 See map of eighteenth century Selborne, between pp. 37 and 38, and Ordnance Survey, 1:50,000.

73 TP i. iv, v, vi, viii, and WJ Ch. Nine. Of all the 'diverse areas' of the old Selborne parish, Woolmer Forest is the most transformed today, with extensive conifer plantations and army training grounds.

74 TP iv.

75 WJ Ch. Nine, and TP i.

76 TP iv. Inside the lime kilns, 'the workmen use sandy loam instead of mortar; the sand of which fluxes, and runs by the intense heat, and so cases over the whole face of the kiln with a strong vitrified coat like glass, that it is well preserved from injuries of weather, and endures thirty or forty years'.

77 NJ 1771, Nov. 8.

78 NJ 1789, Jan. 3.

79 NJ 1789, Jan. 9, and WJ Ch. Nine.

80 NJ 1792, Sept. 2. See map. We find him checking the depths of people's wells on at least one of his journeys along the Sussex Downs; at 'Montham on the down', he found a well which was 'full 350 feet' deep. NJ 1773, Dec. 17 and 18.

81 TP v.

82 NJ 1781, Sept. 26.

83 DB xlv.

84 TP iii.

85 No doubt he got it wrong; according to James Fisher, it was *Ostraea ricordeana. The Natural History of Selborne*, ed. James Fisher (1941), note 4.

86 TP iii.

87 See Daines Barrington, 'On the Deluge in the Time of Noah', in *Miscellanies* (1781).

88 Cf. Robert Plot, *The Natural History of Oxfordshire* (1677), Ch. v.

89 *Rural Rides*, (1824), 1823, Aug. 7.

90 Ray's *Three Physico-Theological Discourses* – which appeared originally in 1692 – dealt respectively with the Creation, the Deluge and the 'final Catastrophe' of the world. Cf. Thomas Burnet as discussed by Stephen Jay Gould, *Questioning the Millennium* (1997), Part One.

91 John Ray, *The Wisdom of God in the Creation* (1691), p. 105. For eighteenth century expositions of Design, see, for example, the Swedish essays included by Stillingfleet in his *Miscellaneous Tracts* (1762).
92 For the comparable case of Robert Hooke, see, for example, A. Rupert Hall, *From Galileo to Newton, 1630–1720*, (1963), Ch. vii.
93 *The Further Correspondence of John Ray*, ed R. W. T. Gunther (1728), p. 72.
94 *The Natural History of Oxfordshire* (1677), Ch. V.
95 For the 'once-joined' continents, see TP xxiv.
96 See *Systema Naturae* (1758), Introduction. By the 1766 edition, the relevant passage had been omitted; but even when Linnaeus allowed that new species might sometimes appear, he assumed that the purity of the 'original' – created – species remained and would remain unimpaired. Although Ray's botany and systematics were by now 'out of date', the *The Wisdom of God in the Creation* was readily available in the mid-eighteenth century; the twelfth edition appeared in 1750.
97 See, for example, H. B. Glass (ed.), *Fore-runners of Darwin 1745–1859* (1959), and Alister Hardy, *The Living Stream* (1965).
98 'Contingent' seems to cause some confusion. As used here and in the following chapter, it is opposed to 'originally created' or 'preordained and thus inevitable'. (It is not opposed to, and may be entirely compatible with, 'adaptive'.)
99 Ernst Mayr, *The Growth of Biological Thought* (1982), Ch. 7.
100 He tried out various 'new' flowers and plants, such as Lucombe's oak and the Nismes radish. In 1790, he diagnosed an interesting example of hybridisation among game birds; see NJ 1790, Oct. 2.
101 For the history of the preformation belief, see A. Rupert Hall, *From Galileo to Newton 1630–1720* (1963), Ch. vii.

Chapter 10: The Problem of Adaptation

1 The nineteenth-century hunters who doubled as animal catchers – the bombastic 'gorilla man', Paul du Chaillu, was an example – were the antithesis of White, but he would have had serious doubts also about Audubon and even Charles Waterton.
2 Quoted by L. S. Hearnshaw, *A Short History of British Psychology, 1840–1940* (1964), Ch. vi.
3 Fair Isle was an example of what Gilbert White called, 'a great rendezvous, and place of observation' for migrants; but the first permanently manned observatory in Britain, which again lagged far behind various Continental countries here, was that on Skokholm (Ronald Lockley), started officially in 1933. Except for the more or less regularly manned Isle of May, it remained the only station of its kind in Britain for more than another decade.
4 The compiling of national plots using questionnaire results worked with

plants but not birds, in Britain. In Belgium, Germany and Sweden, where it was already possible to brief widespread networks of observers, the plotting of bird populations and movements was more successful. See, for example, David Allen, *The Naturalist in Britain* (1979).

5 Daniel McAlpine, *The Botanical Atlas* (1883), Preface.

6 The latter reached their peak in Gosse's book *Omphalos* (1857): this is practically unreadable today (and perhaps was practically unreadable at the time of its publication), but the circumstances in which it was written are described in the useful autobiography, *Father and Son* (1907), by Edmund Gosse. For one of the 'Bridgewater Treatises', see William Kirby, in Bibliography above.

7 *Father and Son* (1907), Ch. vi.

8 NJ 1789, Jan. 4. He identified the bird with the help of *Ray's Willughby*, he says; but considering the measurements he is careful to give, the identification was perhaps faulty. 'The colymbus was of considerable bulk weighing only three drachms short of three pounds averdupois. It measured in length from the bill to the tail (which was very short) two feet; & to the extremities of the toes, four inches more; & the breadth of the wings expanded was 42 inches.' The black throated diver is a more likely candidate: except that the great northern is rather larger, in their *winter plumage* the two species are very similar, and either can appear off the Channel coast in winter.

9 Charles Darwin himself 'learnt Paley's *Evidences*' as a Cambridge undergraduate. For Paley, see Bibliography; also Appendix 5, above.

10 TP xvii. See also TP xx, the remark concerning the buck's head.

11 NJ 1789, Sept. 27.

12 For the origin of the domestic dove, see TP xliv.

13 DB lviii. Dogs raised important questions. The many kinds of domestic dog were races within one species; and in a paper read before the Royal Society in 1787, John Hunter pointed out not only that the wolf, the jackal and the dog were relatives, but that the dog could interbreed with either of the other two. *Philosophical Transactions* (1787), vol. 77, p. 253.

14 TP xxxviii. Pennant had talked of visiting Goodwood, perhaps in the company of Joseph Banks, but this visit did not take place.

15 The female moose measured 16 hands at only two years of age: White adds, 'What a vast tall beast must a full grown stag be!' TP xxxviii.

16 TP xviii and xxx.

17 TP xxii.

18 DB xxvii.

19 DB xxv.

20 DB liv.

21 NJ 1780, July 28: the farmhouse was that at Priory Farm. See also NJ 1788, September 4.

22 RH II, pp. 219–20. The correspondents stop at seven years, perhaps from discomfort: after ten years, and merely on the basis of geometrical progression, the total would have been almost seventy million birds.

23 DB xxxix.
24 NJ 1792, Sept. 23.
25 NJ 1792, Sept. 24.
26 DB lvi.
27 *Physico-Theology,* re-edited (1721), VII : iii.
28 The nightingale, White notes similarly, 'comes as early in cold cutting springs as in mild ones'. NJ 1793, Apr. 12.
29 NJ 1774, Oct. 26.
30 See Eric Tomlin and Robert S. Steiner, *Bird Migration* (1978), Ch. 2, and Jean Dorst, *The Migration of Birds* (1962), Ch. 11.
31 As usual, the information provided by others had to be carefully sifted. He had been surprised to hear from his brother John that among the birds streaming across the Straits of Gibraltar in autumn were examples of 'all the various sorts of hawks and kites', because he had gathered earlier that hawks and kites did not leave 'north Europe' in winter. See DB ix.
32 John Ray, *The Wisdom of God in the Creation* (1691), p. 96.
33 Ibid., pp. 165–166.
34 See, for example, DB xi, and Ch. 11, sect 2, below. In eighteenth century Britain, a 'passion' was actively directing, and a 'propensity' was also felt; ie there was often only a difference of emphasis between the two.
35 Although his 'new migrant', the ring ouzel, was also a thrush, we remember.
36 Just how the term 'congenerous' is to be understood here is not explained. (It may not be quite the same as the more recent 'congeneric'.) The important point, however, is that by 'congenerous animals' or 'congenerous vegetables' White means species with significant similarities, rather than individuals.
37 For example, DB xxi, DB xvi, and NJ 1773, October 2.
38 See, for example, Alan Moorehead, *Darwin and the Beagle* (1971), Ch. x.
39 In *Selborne*, he makes a general statement, but in an uncharacteristically convoluted sentence: 'How diversified are the modes of life not only of incongruous but even of congenerous animals; and yet their specific distinctions are not more various than their propensities.' DB xlviii. (Here, in the traditional manner, 'specific distinctions' appear as the physical appearances and structures of individuals are compared.)
40 The letter concerning the harvest mouse, written in 1767, is used in Selborne, but with the relevant passage modified: TP xii, and see reproduced manuscript page. However, in a different but related discussion, as White traces the domestic dove to its origins in the wild rock dove, he describes a behavioural trait – the willingness of domestic doves to fly off to caverns and 'rocky fastnesses' to breed – as 'worth an hundred arguments' in favour of this ancestral relationship. TP xliv.
41 *The Origin of Species* (1859), Ch. vi. Darwin may well have been influenced by White here. Cf. the case of the earthworms, in Ch. 12, sect. 3, below.

42 *The Voyage of the Beagle*, ed. Leonard Engel (1962), Ch. xvii. See also *The Origin of Species* (1859), Ch. vi.
43 His ambivalence appears in, for example, DB viii and xi.
44 Whether or not Darwin believed in a 'battle for life', an alleged 'war of all against all' had been discussed by Thomas Hobbes (1588–1679); Hobbes, along with the violence of the seventeenth century, was well remembered in eighteenth-century Britain.
45 TP xl.
46 DB xviii
47 NJ 1776, Aug 26. A mutual and beneficial restraint appears, similarly, as magpies are allowed to feed on the ticks they find on the backs and even the heads of sheep, and as they do not peck out the sheeps' eyes. NJ 1776, Nov. 13.
48 See again DB xi.
49 DB xi.
50 DB lviii.
51 See *The Origin of Species* (1859), Chs iv and v. For the 'increasing importance' Darwin's acquired characters were allowed, see also the sixth edition of the *Origin* (1882), the last edited by Darwin.
52 The view touched on here appeared convincingly only with the advent of twentieth century ethology and ecology, but early versions of what is essentially the same idea are to be found in both C. Lloyd Morgan and J. M. Baldwin; see Bibliography, 1896 and 1902.
53 C. H. Waddington, *The Evolution of an Evolutionist* (1975), Ch. 18. The broader notion of 'environment' gives Waddington too a greater validity.
54 For evolution theory as assumed in this section, see any three of the following: Ernst Mayr, *Animal Species and Evolution* (1963), Chs 1 and 16–18; R. C. Lewontin, *The Genetic Basis of Evolutionary Change* (1974); J. C. Barnard, *Animal Behaviour; Ecology and Evolution* (1983); J. Piaget, *Behaviour and Evolution* (1979); J. Maynard Smith and E. Szathmary, *The Origins of Life* (1999); and C. H. Waddington, *The Evolution of an Evolutionist* (1975).
55 *The Origin of Species* (1859), Ch. vi. His brief account of 'sexual selection' is part of Ch. iv.
56 Darwin provides many examples of socially transmitted behaviour in his later book, *The Descent of Man* (1871), but here too he pays little attention to social norms as themselves contributing to the natural selection process.
57 See *The Origin of Species*, Ch. xiv, in even the heavily revised sixth edition, 1882.
58 See, for example, William Kirby, *The Power, Wisdom and Goodness of God as Manifested in the Creation of Animals* (1835), the seventh of the *Bridgewater Treatises*. Darwin himself, we notice, could refer to instincts as 'mental powers'.
59 Nikolaus Tinbergen, *The Study of Instinct*, revised edn (1969),

Introduction.

60 Konrad Lorenz, *On Aggression* (1966), Ch. vi. See also Nikolaus Tinbergen, *The Study of Instinct* (1951), Chs 1 and 4, and W. H. Thorpe, *Learning and Instinct in Animals* (1956), Ch. ii.

61 They are associated with an attempt not to escape from 'fatalism' in biology but to reintroduce it. See, for example, Stephen Jay Gould, *Ever Since Darwin* (1978), Essays 30 and 32. 'Instinct' itself is treated with circumspection today; as an *entity* of some kind, it is unscientific and can be a mere encumbrance. The invalidity of 'instinct' in this sense, however, gives us no reason to reject either *unlearned reaction* or *unlearned readiness to work or learn in given behavioural areas* where these are observable.

62 Some fishes and some birds, for example, have evolved similar patterns of courtship behaviour.

63 For the history of 'imprinting' and 'releasing', see, for example, W. H. Thorpe, *The Origins and Rise of Ethology* (1979), and P. J. B. Slater, *An Introduction to Ethology* (1985).

64 *An Essay on Bird-Mind*, reprinted in *Essays of a Biologist* (1928). Cf. Gilbert White on passionate expression, p. 64, above, and Marion Stamp Dawkins, *Through Our Eyes Only?* (1993), Ch. Five.

65 Jane Goodall, *In the Shadow of Man* (1971), Chs 10 and 15. For Huxley's paper on the great crested grebe, see Bibliography, 1915; and for a more detailed study of ceremonial behaviour and pair-bonding, see Konrad Lorenz, 'Pair-Formation in Ravens', in Heinz Friedrich (ed.), *Man and Animal* (1972). For some remarkable photographs of 'signalling' behaviour, see Niko Tinbergen and Hugh Falcus, *Signals for Survival* (1970).

66 David Lack, *The Life of the Robin* (1943), Chs v and viii. For important earlier work on territory, see H. Eliot Howard, *Territory in Bird Life* (1920), or Edmund Selous, *Realities of Bird Life* (1927), Ch. xiv.

67 Konrad Lorenz, *King Solomon's Ring* (1952), Ch. 11, and Jane Goodall and Hugo van Lawick, *Innocent Killers* (1970), Chs. 2 and 4.

68 See, for example, Ian Carter, *The Red Kite* (Chelmsford, 2001), Ch. 9.

69 Jane Goodall, *In the Shadow of Man* (1971), Ch 16.

70 See Cecilia M. Heyes and Bennet G. Galef, (eds), *Social Learning in Animals* (1996). Also Konrad Lorenz, *King Solomon's Ring* (1952); John T. Bonner, *The Evolution of Culture in Animals* (1980); and Andrew Whiten et al, 'Cultures in Chimpanzees', *Nature*, vol. 399 (1999): p. 692. For the milk bottles, see Fisher and Hinde, Bibliography.

71 See 'The Comparative Study of Behaviour', in Konrad Lorenz and Paul Leyhausen (eds), *Motivation of Human and Animal Behaviour* (1973); also Alec Nesbitt, *Konrad Lorenz* (1976), Ch. 13.

72 See Nikolaus Tinbergen, *The Herring Gull's World* (1953), Chs. 7 and 14. For Tinbergen on comparison, see 'On War and Peace in Animals and Man', in Heinz Friedrich (ed.), *Man and Animal* (1971).

73 Nicholas B. Davies, 'Studying Behavioural Adaptations', reprinted in

M. Stamp Dawkins, T. R. Haliday and R. Dawkins (eds), *The Tinbergen Legacy* (1991).

74 Tinbergen himself dismissed simple or final 'nature or nurture' assertions; see 'On War and Peace in Animals and Man', in Heinz Friedrich (ed.), *Man and Animal* (1971).

75 For a good introduction to the B.T.O, and in general modern co-ordinated recording, see James Fisher/Jim Flegg, *Watching Birds* (1974), Ch. 6.

76 The problem was similar to that of isolating smoking as the principal cause of lung cancer. For a general discussion of 'comparative' methods, see Paul H. Harvey and Mark D. Pagel, *The Comparative Method in Evolutionary Biology* (1991).

77 White's influence on later natural historians is a still largely unexplored topic. It can be noted, though, that such native British observers as Fraser Darling, Ronald Lockley and James Fisher, all of whom knew and appreciated *Selborne,* provided as it were a platform in the U.K. on which the avowed ethologists could build.

78 *An Essay Concerning Human Understanding* (1690), II : xi : 6–11.

79 *The Wisdom of God in the Creation* (1691), p. 42.

80 *The Duty of Mercy and the Sin of Cruelty to Brute Animals*, ed. Richard D. Ryder (1992), Ch. Three. Here – as in Fielding, Smollett or Dr Johnson – a 'rational creature' is a human being.

81 Quoted by W. H. Thorpe, *The Origins and Rise of Ethology* (1979), Ch. 5.

82 See Karl von Frisch, *The Dancing Bees* (1955), Chs. Eleven and Twelve, and, for example, Marian Stamp Dawkins, *Through Our Eyes Only?* (1993), Ch. Three.

83 Irene M. Pepperberg, 'Social Interaction as a Condition for Learning in Avian Species', in Hank Davis and Dianne Balfour (eds), *The Inevitable Bond* (1992). For one of the various attempts to teach apes the use of a human language, see, for example, Francine Patterson and Eugene Linden, *The Education of Koko* (1982).

84 See Bibliography, 1764.

85 DB xiv.

86 See, for example, NJ 1775, June 26 or 1782, July 9.

87 NJ 1780, May 27. Timothy had first arrived at The Wakes in the March of this year: after some false starts he was quite 'awake', and was marching about, by May 2.

88 White subscribes to what today would be called a 'factor' psychology; see also Ch. 11, sect. iii, below. His behaving creatures are *essentially* multi-variate.

89 Among ethologists concerned with evolution, Lack, Thorpe, A. R. Morrison and H. C. Plotkin were or have been prominent – although the last might not accept the title 'ethologist'.

Chapter 11: 'Living Manners'

1 DB xvii.

2 DB iv.

3 The genuine value of White's caution appears not only from a twentieth century non-Lamarckian understanding of the effects of acquired characters; newly weaned rats apparently prefer the food their mother has depended on during lactation – i.e. a mediated habituation of this kind may have been at work in the case of the 'Chinese' puppies.

4 He has nothing to do with the vexed dispute as to whether animals have immortal souls – although this continued throughout his lifetime.

5 He can refer to the 'propensities' not only of animals but of plants, therefore, as he does in DB xli.

6 Quoted by Walter Johnson; WJ Ch. 2.

7 See, for example, Robert A. Hinde, *Ethology,* (1982).

8 Included with DB xli after White's death. But compare with Pope, for example, *An Essay on Man* (1733–4), II, lines 267–80.

9 Mulso begged his friend to bring out his book 'before the appetite for natural history dies down'. For example, RH vol. I, p. 315.

10 He could find the French naturalists 'strangely prolix', but see, for example, PF p. 316 (Brisson), DB xv (Reaumur) and TP xxvi (Buffon). In *Gilbert White and his Records* (1988), Ch. Eight, footnote 11, Paul Foster provides a list of the main authorities cited or otherwise used in *Selborne*; clearly, as he says, White had a working knowledge of many, if not most, of 'the major texts in natural history of his age'.

11 The most famous representative of this psychological tradition was still Hobbes – the most respected was still Locke – but during the first half of the eighteenth century, writers as different as Hume, Hutchison, Pope, Mandeville and Berkeley had all assumed and made use of it. Representatives of the same tradition in Britain were thin on the ground during the nineteenth century, but when, at the beginning of the twentieth, William McDougall was elaborating his 'hormic' psychology, he was by no means without British antecedents.

12 David Hume, *A Treatise of Human Nature* (1739), Introduction.

13 Ibid., Bk I : iv : vi. Here, Hume also likens the human personality to 'a republic or commonwealth'. (Though Hume is often ironic, this is the view he substantiates at length in Bk II of the *Treatise*.)

14 *An Essay on Man* (1733), introductory remarks; and *Moral Essays*, I (1734), lines 122–265.

15 Smollett's last and most good natured novel is made up entirely of 'letters written by' its various, and varied, characters; the eventual result is a contribution to what one of them calls the 'science of men'.

16 *Tom Jones* (1749), Bk. IX, ch. i. Fielding and Smollett were home-grown and 'Lockean', if they were also influenced by various Continental writers. Later in the century, the same was true of James Boswell. In Boswell's account of Dr Johnson he uses 'only' the latter's observations, written or

spoken; he tries to avoid 'melting down my materials into one mass, and constantly speaking in my own person'. He allows the relevant personality profile to emerge, in the course of his book. See *The Life of Samuel Johnson* (1791), Introductory.

17 *Essay on Man*, II, lines 1 and 2.

18 David Hume, Of National Characters, *Essays Moral, Political and Literary*, XXI (1963); cf. *A Treatise of Human Nature* (1739), Bk III : ii : ii.

19 Joseph Banks was among the foreign travellers bringing back accounts of 'primitive' peoples; he did so most impressively on his return from the circumnavigation with Cook, in 1771. For a letter from William Sheffield giving White details of the large collection of botanical specimens, but also garments, weapons, tools and other artefacts, put on display by Banks in 1772, see, for example, Patrick O'Brian, *Joseph Banks* (1987), Ch. 7.

20 White's willingness to recognise emotion perhaps links him with his immediate contemporaries, but his is a dispassionate description of 'passion'; as I have insisted, he is closer to Pope, Fielding or Hume himself. See eg Hume, *Treatise of Human Nature*, Bk II.

21 RH II, Ch. xli.

22 Here too White was a good Lockean. Locke recognised an immediate and intuitive movement of the mind, if he also insisted on careful verification. See *An Essay Concerning Human Understanding* (1690), Bk IV, ii, 1–7.

23 DB xviii.

24 E. A. R. Ennion, *The British Bird* (1943), Foreword.

25 Even the act of putting the tortoise in a tub of water to find out whether it could swim, NJ 1780, July 1, though it revealed something, was hardly experimental. (Perhaps in the wild, tortoises of this species learn to swim.)

26 This is not to say that Hales did not influence White here. Hales's view that more dew falls on a wet than on a dry surface seems to have been backed up by controlled experiments: he used shallow pans of wet and dry soil, and apparently weighed these carefully before and after dew had fallen on them. See *Vegetable Staticks* (1733), Ch. I, exp. xix. Hales lacks his usual clarity on this occasion; but White could respond creatively to a passing remark.

27 Copies were available at the end of 1788, but the book was reviewed only in the early months of 'MDCCLXXXIX', the date on the title page.

28 DB xlix. Cf. TP xv (albino rooks) and TP xxxix (osprey). By the end of this 'enlightened' century in Britain, we remember, upwards of two hundred human offences were punishable by death.

29 RH vol. II, p. 256. Even in *Selborne* he refers to 'unreasonable sportsmen', and remarks coolly that the end of a barn is 'the countryman's museum'. TP vi and x.

30 See NJ 1784, Oct. 16 (inserted newspaper cutting).

31 Thinking about field crickets, he can temporarily make himself the same

size as a field cricket; he refers, for instance, to this creature's 'vast row of serated fangs'. DB xlvi.

32 TP xvi.

33 This 'coda', originally added to AN xxvi, is in some editions included in *Selborne* itself, with DB l. This 'makes sense' – but the sly humour of the original positioning is lost.

34 Even the poem *The Naturalist's Summer-Evening Walk,* which was included in *Selborne* by White himself, should be seen as a graceful but principally decorative addition.

35 RH vol. II, p. 159. The higher parts of the South Downs were perhaps another matter: see, for example, DB xvii.

36 See the exchanges between Gilbert and Molly (Mary) White in RH vol. II.

37 The use of equipment in work on live animals was entirely conceivable in White's time, we can notice again. There were the vivisectionists, and Barrington was conducting his investigations with song-birds. Max Nicholson refers to a serious, if inconclusive, attempt on the part of John Hunter to study the alleged hibernation of the hirundine species by leaving some swallows in a closed room during the autumn, a room in which various 'suitable hibernation sites', such as hollow logs and receptacles filled with water, had been arranged. (The swallows died.) *The Natural History of Selborne,* ed. E. M. Nicholson (1929), Introduction.

38 Pope – himself an anti-vivesectionist – seems to suggest that it is impossible to analyze the 'quick whirls and shifting eddies' of a mind or life, human or non-human; but he goes on to disprove this in the same poem. *Moral Essays*, I (1934), lines 29–40.

39 Stephen Jay Gould, *Ever Since Darwin* (1978), Epilogue.

Chapter 12: The Perennial Naturalist

1 Critics are asked to read this chapter before judging it.

2 Most scientists of the last hundred years have been concerned with whether given claims as to what happens in the existential world survive experimental testing, rather than with whether the objects and mechanisms in question are 'really there'. Pushed to the metaphysical point, most of these scientists would have been neither 'nomenalists' nor 'realists' – but something of both.

3 Thomas S. Kuhn, *The Structures of Scientific Revolutions* (1970), Postscript.

4 The famous case of the Soviet biologist T. D. Lysenko shows not that science is 'merely cultural', but that, even in the modern age, the scientist who cooks his books is likely sooner or later to be exposed.

5 The so-called *Historical Sketch,* included by, for example, J. W. Burrow in the Penguin edition of *The Origin of Species* (1982).

6 See, for example, P. Bridgeman, *The Logic of Modern Physics* (1948), and David M. Knight, *The Nature of Science* (1976). This too may be only

a temporary state of affairs; 'metaphysical science' constantly threatens 'phenomenological' science, as the British empiricists saw. The continuing relevance of the empiricists is none the less striking. While discussing the ascertaining of causes, Hume remarks: 'There is no phaenomenon in nature, but what is compounded and modify'd by so many different circumstances, that in order to arrive at the decisive point, we must carefully separate whatever is superfluous, and enquire by new experiments, if every particular circumstance of the first experiment was essential to it.' *A Treatise of Human Nature* (1739), I : iii : xv.

7 See, for example, *On Aggression* (1966), Ch. vi, and Alec Nesbitt, *Konrad Lorenz* (1976), Ch. 13.

8 *The Study of Instinct* (1951), Introduction.

9 See Marian Stamp Dawkins, *Through Our Eyes Only? The Search for Animal Consciousness* (1993).

10 *The Natural History of Selborne*, ed. James Fisher (1941), Preface.

11 He gives us the exact measurements even of his 'huge' hail stones.

12 DB xxxv.

13 Charles Darwin, *The Formation of Vegetable Mould Through the Action of Worms* (1881), Ch. 1.

INDEX

Works cited only in the Bibliography or the Notes and References section are not included here. Scientific or 'Latin' names of species, many of which have been changed since White's time, are not included, and modern spellings – without capitals – are given for the English names.